DATE DUE   8

Demco

8

Musculoskeletal Models and Techniques

BIOMECHANICAL SYSTEMS
TECHNIQUES AND APPLICATIONS

VOLUME III

EDITED BY

CORNELIUS LEONDES

CRC Press
Boca Raton London New York Washington, D.C.

Library of Congress Cataloging-in-Publication Data

Catalog record is available from the Library of Congress.

© 2001 by CRC Press LLC

No claim to original U.S. Government works
International Standard Book Number 0-8493-9048-6
Printed in the United States of America 1 2 3 4 5 6 7 8 9 0
Printed on acid-free paper

Preface

Because of rapid developments in computer technology and computational techniques, advances in a wide spectrum of technologies, and other advances coupled with cross-disciplinary pursuits between technology and its applications to human body processes, the field of biomechanics continues to evolve. Many areas of significant progress can be noted. These include dynamics of musculoskeletal systems, mechanics of hard and soft tissues, mechanics of bone remodeling, mechanics of implant-tissue interfaces, cardiovascular and respiratory biomechanics, mechanics of blood and air flow, flow-prosthesis interfaces, mechanics of impact, dynamics of man–machine interaction, and more.

Needless to say, the great breadth and significance of the field on the international scene require several volumes for an adequate treatment. This is the third in a set of four volumes, and it treats the area of musculoskeletal models and techniques.

The four volumes constitute an integrated set that can nevertheless be utilized as individual volumes. The titles for each volume are

Computer Techniques and Computational Methods in Biomechanics
Cardiovascular Techniques
Musculoskeletal Models and Techniques
Biofluid Methods in Vascular and Pulmonary Systems

The contributions to this volume clearly reveal the effectiveness and significance of the techniques available and, with further development, the essential role that they will play in the future. I hope that students, research workers, practitioners, computer scientists, and others on the international scene will find this set of volumes to be a unique and significant reference source for years to come.

The Editor

Cornelius T. Leondes, B.S., M.S., Ph.D., Emeritus Professor, School of Engineering and Applied Science, University of California, Los Angeles has served as a member or consultant on numerous national technical and scientific advisory boards. Dr. Leondes served as a consultant for numerous Fortune 500 companies and international corporations. He has published over 200 technical journal articles and has edited and/or co-authored more than 120 books. Dr. Leondes is a Guggenheim Fellow, Fulbright Research Scholar, and IEEE Fellow as well as a recipient of the IEEE Baker Prize award and the Barry Carlton Award of the IEEE.

Contributors

Eihab Muhammed Abdel-Rahman, Ph.D.
Virginia Polytechnic Institute
Blacksburg, VA

Ali E. Engin, Ph.D.
University of South Alabama
Mobile, AL

Vijay K. Goel, Ph.D.
University of Iowa
Iowa City, IA

Nicole M. Grosland, B.S.E.
University of Iowa
Iowa City, IA

Robert D. Harten, Jr., Ph.D.
New Jersey Medical School
Newark, NJ

David Hawkins, Ph.D.
University of California at Davis
Davis, CA

Mohamed Samir Hefzy, Ph.D., P.E.
The University of Toledo
Toledo, OH

J. Lawrence Katz, Ph.D.
Case Western Reserve University
Cleveland, OH

Phyllis Kristal
Harborview Medical Center
Seattle, WA

Roderic S. Lakes, Ph.D.
University of Wisconsin at Madison
Madison, WI

Alain Meunier, Ph.D.
L'Hopital St. Louis
Paris, France

Chimba Mkandawire, B.Sc.
Harborview Medical Center
Seattle, WA

Barry S. Myers, M.D., Ph.D.
Duke University
Durham, NC

Sheu-Jane Shieh, Ph.D.
Wayne State University
Detroit, MI

Allan F. Tencer, Ph.D.
Harborview Medical Center
Seattle, WA

Chris Van Ee, Ph.D.
University of Michigan
Ann Arbor, MI

Mark C. Zimmerman, Ph.D.
Johnson & Johnson Corp.
Somerville, NJ

Contents

1

Three-Dimensional Dynamic Anatomical Modeling of the Human Knee Joint

Mohamed Samir Hefzy
University of Toledo

Eihab Muhammed Abdel-Rahman
Virginia Polytechnic Institute

Three-dimensional dynamic anatomical modeling of the human musculo-skeletal joints is a versatile tool for the study of the internal forces in these joints and their behavior under different loading conditions following ligamentous injuries and different reconstruction procedures. This chapter describes the three-dimensional dynamic response of the tibio-femoral joint when subjected to sudden external pulsing loads utilizing an anatomical dynamic knee model. The model consists of two body segments in contact (the femur and the tibia) executing a general three-dimensional dynamic motion within the constraints of the ligamentous structures. Each of the articular surfaces at the tibio-femoral joint is represented by a separate mathematical function. The joint ligaments are modeled as nonlinear elastic springs. The six-degrees-of-freedom joint motions are characterized using six kinematic parameters and ligamentous forces are expressed in terms of these six parameters. In this formulation, all the coordinates of the ligamentous attachment points are dependent variables which allow one to easily introduce more ligaments and/or split each ligament into several fiber bundles. Model equations consist of nonlinear second order ordinary differential equations coupled with nonlinear algebraic constraints. An algorithm was developed to solve this differential-algebraic equation (DAE) system by employing a DAE solver, namely, the Differential Algebraic System Solver (DASSL) developed at Lawrence Livermore National Laboratory.

Model calculations show that as the knee was flexed from 15 to 90°, it underwent internal tibial rotation. However, in the first 15° of knee flexion, this trend was reversed: the tibia rotated internally as the knee was extended from 15° to full extension. This indicates that the screw-home mechanism that calls for external rotation in the final stages of knee extension was not predicted by this model. This finding is important since it is in agreement with the emerging thinking about the need to re-evaluate this mechanism.

It was also found that increasing the pulse amplitude and duration of the applied load caused a decrease in the magnitude of the tibio-femoral contact force at a given flexion angle. These results suggest that increasing load level caused a decrease in joint stiffness. On the other hand, increasing pulse amplitude did not change the load sharing relations between the different ligamentous structures. This was expected since the forces in a ligament depend on its length which is a function of the relative position of the tibia with respect to the femur.

Reciprocal load patterns were found in the anterior and posterior fibers of both anterior and posterior cruciate ligaments, (ACL) and (PCL), respectively. The anterior fibers of the ACL were slack at full extension and tightened progressively as the knee was flexed, reaching a maximum at 90° of knee flexion. The posterior fibers of the ACL were most taut at full extension; this tension decreased until it vanished around 75° of knee flexion. The forces in the anterior fibers of the PCL increased from zero at full extension to a maximum around 60° of knee flexion, and then decreased to 90° of knee flexion. On the other hand, the posterior fibers of the PCL were found to carry lower loads over a small range of motion; these forces were maximum at full extension and reached zero around 10° of knee flexion. These results suggest that regaining stability of an ACL deficient knee would require the reconstruction of both the anterior and posterior fibers of the ACL. On the other hand, these data suggest that it might be sufficient to reconstruct the anterior fibers of the PCL to regain stability of a PCL deficient knee.

1.1   Background

Biomechanical Systems

Biomechanics is the study of the structure and function of biological systems by the means of the methods of mechanics.[63] Biomechanics thus provides the means to study and analyze the behaviors of the different biological systems as well as their components. Models have emerged as necessary and effective tools to be employed in the analysis of these biomechanical systems.

In general, employing models in system analysis requires two prerequisites: a clear objective identifying the aims of the study, and an explicit specification of the assumptions to be made. A system mainly depends upon what it is being used to determine, e.g., joint stiffness or individual ligament lengths and forces. The system is thus identified according to the aim of the study. Assumptions are then introduced in order to simplify the system and construct the model. These assumptions also depend on the aim of the study. For example, if the intent is to determine the failure modes of a tendon, it is not reasonable to model the action of a muscle as a single force applied to the muscle's attachement point. On the other hand, this assumption is appropriate if it is desired to determine the effect of a tendon transfer on gait.

After the system has been defined and simplified, the modeling process continues by identifying system variables and parameters. The parameters of a system characterize its components while the variables describe its response. The variables of a system are also referred to as the quantities being determined.

Modeling activities include the development of physical models and/or mathematical models. The mechanical responses of physical models are determined by conducting experimental studies on fabricated structures to simulate some aspect of the real system. Mathematical models satisfy some physical laws and consist of a set of mathematical relations between the system variables and parameters along with a solution method. These relations satisfy the boundary and initial conditions, and the geometric constraints.

Major problems can be encountered when solving the mathematical relations forming a mathematical model. They include indeterminacy, nonlinearity, stability, and convergence. Several procedures have been developed to overcome some of these difficulties. For example, eliminating indeterminacy requires using linear or nonlinear optimization techniques. Different numerical algorithms[16] presented in the literature effectively allow solution of most nonlinear systems of equations, yet reaching a stable and convergent solution never may be achieved in some situations. However, with the recent advances in computers, mathematical models have proved to be effective tools for understanding behaviors of various components of the human musculo-skeletal system.

Mathematical models are practical and appealing because:

1. For ethical reasons, it is necessary to test hypotheses on the functioning of the different components of the musculo-skeletal system using mathematical simulation before undertaking experimental studies.[115]
2. It is more economical to use mathematical modeling to simulate and predict joint response under different loading conditions than costly *in vivo* and/or *in vitro* experimental procedures.
3. The complex anatomy of the joints means it is prohibitively complicated to instrument them or study the isolated behaviors of their various components.
4. Due to the lack of noninvasive techniques to conduct *in vivo* experiments, most experimental work is done *in vitro*.

This chapter focuses on the human knee joint which is one of the largest and most complex joints forming the musculoskeletal system. From a mechanical point of view, the knee can be considered as a biomechanical system that comprises two joints: the tibio-femoral and the patello-femoral joints. The behavior of this complicated system largely depends on the characteristics of its different components. As indicated above, models can be physical or mathematical. Review of the literature reveals that few physical models have been constructed to study the knee joint. Since this book is concerned with techniques developed to study different biomechanical systems, the few physical knee models will be discussed briefly in this background section.

Physical Knee Models

Physical models have been developed to determine the contact behavior at the articular surfaces and/or to simulate joint kinematics. In order to analyze the stresses in the contact region of the tibio-femoral joint, photoelasticity techniques have been employed in which epoxy resin was used to construct models of the femur and tibia.[31,87] Kinematic physical knee models were also proposed to demonstrate the complex tibio-femoral motions that can be described as a combination of rolling and gliding.[100,101]

The most common physical model that has been developed to illustrate the tibio-femoral motions is the crossed four-bar linkage.[76,88,89] This construct consists of two crossed rods that are hinged at one end and have a length ratio equal to that of the normal anterior and posterior cruciate ligaments. The free ends of these two crossed rods are connected by a coupler that represents the tibial plateau. This simple apparatus was used to demonstrate the shift of the contact points along the tibio-femoral articular surfaces that occur during knee flexion.

Another model, the Burmester curve, has been used to idealize the collateral ligaments.[90] This curve is comprised of two third order curves: the vertex cubic and the pivot cubic. The construct combining the crossed four-bar linkage and the Burmester curve has been used extensively to gain an insight into knee function since the cruciate and collateral ligaments form the foundation of knee kinematics. However, this model is limited because it is two-dimensional and does not bring tibial rotations into the picture. A three-dimensional model proposed by Huson allows for this additional rotational degree-of-freedom.[76]

Phenomenological Mathematical Knee Models

Several mathematical formulations have been proposed to model the response of the knee joint which constitutes a biomechanical system. Three survey papers appeared in the last decade to review mathematical knee models which can be classified into two types: *phenomenological* and *anatomically based* models.[66,69,70]

The phenomenological models are gross models, describing the overall response of the knee without considering its real substructures. In a sense, these models are not real knee models since a model's effectiveness in the prediction of *in vivo* response depends on the proper simulation of the knee's articulating surfaces and ligamentous structures. Phenomenological models are further classified into *simple hinge* models, which consider the knee a hinge joint connecting the femur and tibia, and *rheological* models, which consider the knee a viscoelastic joint.

Simple Hinge Models

This type of knee model is typically incorporated into global body models. Such whole-body models represent body segments as rigid links connected at the joints which actively control their positions. Some of these models are used to calculate the contact forces in the joints and the muscle load sharing during specific body motions such as walking,[38,73,86,112,114] running,[25] and lifting and lowering tasks.[39,44] These models provide no details about the geometry and material properties of the articular surfaces and ligaments. Equations of motion are written at the joint and an optimization technique is used to solve the system of equations for the unknown muscle and contact forces. Other simple hinge models were developed to predict impulsive reaction forces and moments in the knee joint under the impact of a kick to the leg in the sagittal plane.[83,121,122] In these models, the thigh and the leg were considered as a double pendulum and the impulse load was expressed as a function of the initial and final velocities of the leg.

Rheological Models

These models use linear viscoelasticity theory to model the knee joint using a Maxwell fluid approximation[97] or a Kelvin body idealization.[37,108] Masses, springs, and dampers are used to represent the velocity-dependent dissipative properties of the muscles, tendons, and soft tissue at the knee joint. These models do not represent the behavior of the individual components of the knee; they use experimental data to determine the overall properties of the knee. While phenomenological models are of limited use, their dynamic nature makes them of interest.

Anatomically Based Mathematical Knee Models

Anatomically based models are developed to study the behaviors of the various structural components forming the knee joint. These models require accurate description of the geometry and material properties of knee components. The degree of sophistication and complexity of these models varies as rigid or deformable bodies are employed. The analysis conducted in most of the knee models employs a system of rigid bodies that provides a first order approximation of the behaviors of the contacting surfaces. Deformable bodies have been introduced to allow for a better description of this contact problem.

Employing rigid or deformable bodies to describe the three-dimensional surface motions of the tibia and/or the patella with respect to the femur using a mathematical model requires the development of a three-dimensional mathematical representation of the articular surfaces. Methods include describing the articular surfaces using a combination of geometric primitives such as spheres, cones, and cylinders,[4-7,116,125,136,137] describing each of the articular surfaces by a separate polynomial function of the form $y = y (x, z)$,[21,23,75] and describing the articular surfaces utilizing the piecewise continuous parametric bicubic Coons patches.[12,14,67,68] The B-spline least squares surface fitting method is also used to create such geometric models.[13]

Hefzy and Grood[66] further classified anatomically based models into *kinematic* and *kinetic* models. Kinematic models describe and establish relations between motion parameters of the knee joint. They do not, however, relate these motion parameters to the loading conditions. Since the knee is a highly compliant structure, the relations between motion parameters are heavily dependent on loading conditions making each of these models valid only under a specific loading condition. Kinetic models try to remedy this problem by relating the knee's motion parameters to its loading condition.

In turn kinetic models are classified as *quasi-static* and *dynamic*. Quasi-static models determine forces and motion parameters of the knee joint through solution of the equilibrium equations, subject to appropriate constraints, at a specific knee position. This procedure is repeated at other positions to cover a range of knee motion. Quasi-static models are unable to predict the effects of dynamic inertial loads which occur in many locomotor activities; as a result, dynamic models have been developed. Dynamic models solve the differential equations of motion, subject to relevant constraints, to obtain the forces and motion parameters of the knee joint under dynamic loading conditions. In a sense, quasi-static models march on a space parameter, for example, flexion angle, while dynamic models march on time.

Quasi-Static Anatomically Based Knee Models

Several three-dimensional anatomical quasi-static models are cited in the literature. Some of these models are for the tibio-femoral joint, some for the patello-femoral joint, some include both tibio-femoral and patello-femoral joints, and some include the menisci. The most comprehensive quasi-static models for the tibio-femoral joint include those developed by Wismans et al.,[129,130] Andriacchi et al.,[9] and Blankevoort et al.[20-23] The most comprehensive quasi-static three-dimensional models for the patello-femoral joint include those developed by Heegard et al.,[64] Essinger et al.,[50] Hirokawa,[72] and Hefzy and Yang.[68] The models developed by Tumer and Engin,[118] Gill and O'Connor,[57] and Bendjaballah et al.[17] are the only models that realistically include both tibio-femoral and patello-femoral joints. The latter model is the only and most comprehensive quasi-static three-dimensional model of the knee joint available in the literature.[17] This model includes menisci, tibial, femoral and patellar cartilage layers, and ligamentous structures. The bony parts were modeled as rigid bodies. The menisci were modeled as a composite of a matrix reinforced by collagen fibers in both radial and circumferential directions. However, this comprehensive model is limited because it is valid only for one position of the knee joint: full extension.

This chapter is devoted to the dynamic modeling of the knee joint. Therefore, the previously cited quasi-static models will not be further discussed. The reader is referred to the review papers on knee models by Hefzy et al. for more details on these quasi-static models.[66,70]

Dynamic Anatomically Based Knee Models

Most of the dynamic anatomical models of the knee available in the literature are two-dimensional, considering only motions in the sagittal plane. These models are described by Moeinzadeh et al.,[93-99] Engin and Moeinzadeh,[47] Wongchaisuwat et al.,[131] Tumer et al.,[118-119] Abdel-Rahman and Hefzy,[1-3] and Ling et al.[84]

Moeinzadeh et al.'s two-dimensional model of the tibio-femoral joint represented the femur and the tibia by two rigid bodies with the femur fixed and the tibia undergoing planar motion in the sagittal plane.[93,96] Four ligaments, the two cruciates and the two collaterals were modeled by a spring element each. Ligamentous elements were assumed to carry a force only if their current lengths were longer than their initial lengths, which were determined when the tibia was positioned at 54.79° of knee flexion. A quadratic force elongation relationship was used to calculate the forces in the ligamentous elements. A one contact point analysis was conducted where normals to the surfaces of the femur and the tibia, at the point of contact, were considered colinear. The profiles of the femoral and tibial articular surfaces were measured from X-rays using a two-dimensional sonic digitizing technique. A polynomial equation was generated as an approximate mathematical representation of the profile of each surface. Results were presented for a range of motion from 54.79° of knee flexion to full extension under rectangular and exponential sinusoidal decaying forcing pulses passing through the tibial center of mass. No external moments were considered in the numerical calculation.

Moeinzadeh et al.'s theoretical formulation included three differential equations describing planar motion of the tibia with respect to the femur, and three algebraic equations describing the contact condition and the geometric compatibility of the problem.[93-96] Using Newmark's constant-average-acceleration scheme,[15] the three differential equations of motion were transformed to three nonlinear algebraic equations. Thus, the system was reduced to six nonlinear algebraic equations in six independent unknowns: the x and y coordinates of the origin of the tibial coordinate system with respect to the femoral system, the angle of knee flexion, the magnitude of the contact force, and the x coordinates of the contact point in both the femoral and tibial coordinate systems. However, instead of using the differential form of the Newton-Raphson iteration technique to solve these six nonlinear algebraic equations in their numerical analysis, Moeinzadeh et al. used an incremental form of the Newton-Raphson technique. Thus, they reformulated the system of equations to include 22 equations in 22 unknowns. This system was solved iteratively. In this formulation they considered the coordinates of the ligamentous tibial insertion sites (moving points) as eight independent variables and added eight compatibility equations for the locations of these ligamentous tibial insertion sites. The remaining independent variables included

1. The x and y components of the unit vectors normal to the femoral and to the tibial profiles at the point of contact (four variables)
2. The y coordinates of the contact point in both the femoral and tibial coordinate systems (two variables)
3. The slope of the articular profiles at the contact point expressed in both femoral and tibial coordinate systems (two variables)

Moeinzadeh et al.'s model was limited since it was valid only for a range of 0° to 55° of knee flexion. This limitation was a result of their mathematical representation of the femoral profile that diverged significantly from the anatomical one in the posterior part of the femur and their assumption that all ligaments were only taut at 54.79° of knee flexion.

Moeinzadeh et al. extended their work and presented a formulation for the three-dimensional version of their model. However, they were not able to obtain a solution because of "... the extreme complexity of the equations." Their solution technique required them to consider an additional 85 variables as independent and add 85 compatibility equations to solve a system of 101 equations in 101 unknowns.

Wongchaisuwat et al.[131] presented a dynamic model to analyze the planar motion between the femoral and tibial contact surfaces in the sagittal plane. In their model, the authors considered the tibia as a pendulum that swings about the femur. Newton's and Euler's equations of motion were then used to formulate the gliding and rolling motions defined by holonomic and nonholonomic conditions, respectively. Using their model, the authors presented a control strategy to cause the motion and maintain the contact between the surfaces. Their control system included two classes of input: muscle forces, which caused and stabilized the motion, and ligament forces, which maintained the constraints.

To investigate the applicability of classical impact theory to an anatomically based model of the tibiofemoral joint, Engin and Tumer[48,49] developed a modified version of Moeinzadeh et al.'s model. Unstrained lengths of the ligaments were calculated by assuming strain levels at full extension. The model used a two-piece force-elongation relationship, including linear and quadratic regions, to evaluate the ligamentous forces. Engin and Tumer proposed two improved methods to obtain the response of the knee joint using this model. These are the minimal differential equation (MDE) and the excess differential equation (EDE) methods.

In the MDE method, the algebraic equations (constraints) are eliminated through their use to express some variables in terms of others in closed form. Furthermore, one of the differential equations of motion is used to express the contact force in terms of the other variables. It is then used in the other differential equations to eliminate the contact force from the differential equations system, thus reducing that system by one equation. The resulting nonlinear ordinary differential equation system is then solved using both Euler and Runge-Kutta methods of numerical integration.

In the EDE method, the algebraic constraints are converted to differential equations by differentiating them twice with respect to time, producing a second order ordinary differential equation system in the position parameters (five variables). One equation of motion is dropped from the system of equations and used to express the magnitude of the contact force in terms of the other variables. The system is thus reduced to a system of five differential equations in five unknowns. This system of equations is then integrated numerically using both Euler and Runge-Kutta methods of numerical integration. Upon evaluating the position parameters, the last equation of motion is solved for the contact force. The basic assumption in this method is that if the constraints are satisfied initially, then satisfying the second derivatives of the constraints in future time steps is expected to satisfy the constraints themselves.

Tumer and Engin[118] extended the Engin and Tumer model[48,49] to include both the tibio-femoral and the patello-femoral joints and introduced a two-dimensional, three-body segment dynamic model of the knee joint. The model incorporated the patella as a massless body and the patellar ligament as an inextensible link. At each time step of the numerical integration, the system of equations governing the tibio-femoral joint was solved using the MDE method, then the system of equations governing the motion of the patello-femoral joint, a non-linear algebraic equations system, was solved using the Newton-Raphson method.

Abdel-Rahman and Hefzy presented a modified version of Moeinzadeh et al.'s model.[1-4] A part of a circle was used to represent the profile of the femur and a parabolic polynomial was used to represent the tibia. Ten ligamentous elements were used to model the major knee ligaments and the posterior fibers of the capsule. The unstrained lengths of the ligamentous elements were calculated by assuming strain levels at full extension. A quadratic force elongation relationship was used to evaluate the ligamentous forces. Results were obtained for knee motions under a sudden impact simulated by a posterior forcing pulse in the form of a rectangular step function applied to the tibial center of gravity when the knee joint was at full extension; knee motions were tracked until 90° knee flexion was achieved. The results demonstrated the effects of varying the pulse amplitude and duration on the velocity and acceleration of the tibia, as well as on the magnitude of the contact force and on the different ligamentous forces.

Furthermore, Abdel-Rahman and Hefzy introduced another approach, the reverse EDE method, to solve the two-dimensional dynamic model of the tibio-femoral joint.[1-4] In the reverse EDE method, the Newmark method is used to transform the differential equations of motion into non-linear algebraic equations. Combining these equations with the non-linear algebraic constraints, the resulting nonlinear algebraic system of equations is solved using the differential form of the Newton-Raphson method.

In Moeinzadeh et al.'s formulation, the coordinates of the ligaments' insertion sites were considered as independent variables. This approach caused the model to become more complicated when more ligaments were introduced or existing ligaments were subdivided into several elements. This major problem was solved by the Abdel-Rahman and Hefzy formulation in which all the coordinates of the ligaments' insertion sites were considered as dependent variables. As a result, introducing more ligaments to the model or splitting existing ligaments into several fiber bundles to better represent them did not affect the system to be solved. Furthermore, Abdel-Rahman and Hefzy used a more anatomical femoral profile, enabling them to predict tibio-femoral response over a range of motion from 0 to 90° of knee flexion.[1-3]

In summary, since Wongchaisuwat et al.'s model[131] is more a control strategy to cause and maintain contact between the femur and the tibia, it is not considered a real mathematical model that predicts knee response under dynamic loading. Most of the remaining dynamic models[1-3,47-49,93-96] can be perceived as different versions of a single dynamic model. Such a model is comprised of two rigid bodies: a fixed femur and a moving tibia connected by ligamentous elements and having contact at a single point. The various versions of this model have severe limitations in that they are two-dimensional in nature. A three-dimensional dynamic version of the model was presented by Moeinzadeh and Engin.[98] However, obtaining results using their formulation was not possible because of the limitations of the solution technique.

In this chapter, we present the three-dimensional version of this dynamic model. A new approach, the modified reverse EDE method is presented and used to solve the governing system of equations. In this solution technique, the second order time derivatives are first transformed to first order time derivatives then they are combined with the algebraic constraints to produce a system of differential algebraic equations (DAEs). The DAE system is solved using a DAE solver, namely, the differential/algebraic system solver (DASSL) developed at Lawrence Livermore National Laboratory. This DAE solver will be discussed in detail. Model calculations will be presented for exponentially decaying sinusoidal forcing pulses with different amplitudes and time durations. Results will be reported to describe the knee response including the medial and lateral contact pathways on both femur and tibia, the medial and lateral contact forces, and the ligamentous forces. A comparison of model predictions with the limited experimental data available in the literature will then be presented. Finally, a discussion on how this dynamic three-dimensional knee model can be further developed to incorporate the patello-femoral joint will be included.

1.2 Three-Dimensional Dynamic Modeling of the Tibio-Femoral Joint: Model Formulation

The femur and tibia are modeled as two rigid bodies. Cartilage deformation is assumed relatively small compared to joint motions[129-130] and not to affect relative motions and forces within the tibio-femoral joint. Furthermore, friction forces will be neglected because of the extremely low coefficients of friction of the articular surfaces.[99,110] Hence, in this model, the resistance to motion is essentially due to the ligamentous structures and the contact forces. Nonlinear spring elements were used to simulate the ligamentous structures whose functional ranges are determined by finding how their lengths change during motion. The menisci were not taken into consideration in the present model. The rationale is that loading conditions will be limited to those where the knee joint is not subjected to external axial compressive loads. This is based on the numerous reports in the literature indicating that the effect of meniscectomy on joint motions is minimal compared to that of cutting ligaments in the absence of joint axial compressive loads.[113,129]

Kinematic Analysis

Six quantities are used to fully describe the relative motions between moving and fixed rigid bodies: three rotations and three translations. These rotations and translations are the components of the rotation and translation vectors, respectively. The three rotation components describe the orientation of the moving system of axes (attached to the moving rigid body) with respect to the fixed system of axes (attached to the fixed rigid body). The three translation components describe the location of the origin of the moving system of axes with respect to the fixed one.

The tibio-femoral joint coordinate system introduced by Grood and Suntay was used to define the rotation and translation vectors that describe the three-dimensional tibio-femoral motions.[61] This joint coordinate system is shown in Fig. 1.1 and consists of an axis (x-axis) that is fixed on the femur ($\hat{i}$ is a unit vector parallel to the x-axis), an axis (z′-axis) that is fixed on the tibia ($\hat{k}'$ parallel to the z′-axis), and a floating axis perpendicular to these two fixed axes ($\hat{e}_2$ is a unit vector parallel to the floating axis). The three components of the rotation vector include flexion-extension, tibial internal-external, and varus-valgus rotations. Flexion-extension rotations, α, occur around the femoral fixed axis; internal-external tibial rotations, γ, occur about the tibial fixed axis; and varus-valgus rotations, β, (ad-abduction) occur about the floating axis. Using this joint coordinate system, the rotation vector, $\vec{\theta}$, describing the orientation of the tibial coordinate system with respect to the femoral coordinate system is written as:

$$\vec{\theta} = -\alpha\,\hat{i} - \beta\,\hat{e}_2 - \gamma\,\hat{k}' \tag{1.1}$$

This rotation vector can be transformed to the femoral coordinate system, then differentiated with respect to time to yield the angular velocity and angular acceleration vectors of the tibia with respect to the femur.

In this analysis, it is assumed that the femur is fixed while the tibia is moving. The locations of the attachment points of the ligamentous structures as well as other bony landmarks are specified on each bone and expressed with respect to a local bony coordinate system. The distances between the tibial and femoral attachment points of the ligamentous structures are calculated in order to determine how the lengths of the ligaments change during motion. Analysis includes expressing the coordinates of each attachment point with respect to one bony coordinate system: the tibia or the femur. This is accomplished by establishing the transformation between the two coordinate systems. The six parameters (three rotations and three translations) describing tibio-femoral motions were used to determine this transformation as follows:

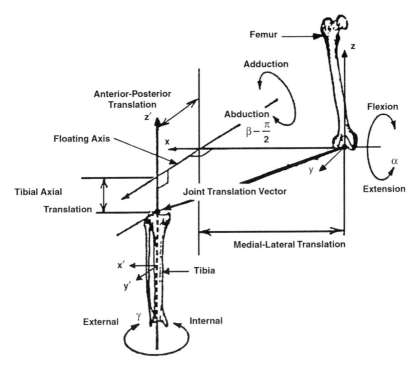

FIGURE 1.1   Tibio-femoral joint coordinate system. (*Source*: Rahman, E.M. and Hefzy, M.S., *JJBE: Med. Eng. Physics*, 20, 4, 276, 1998. With permission from Elsevier Science.)

$$\vec{R} = \vec{R}_o + [R]\vec{r} \tag{1.2}$$

where $\vec{r}$ describes the position vector of a point with respect to the tibial coordinate system, and $\vec{R}$ describes the position vector of the same point with respect to the femoral coordinate system. The vector $\vec{R}_o$ is the position vector which locates the origin of the tibial coordinate system with respect to the femoral coordinate system, and [R] is a (3 × 3) rotation matrix given by Grood and Suntay[61] as:

$$[R] = \begin{bmatrix} \sin\beta\cos\gamma & \sin\beta\sin\gamma & \cos\beta \\ -\cos\alpha\sin\gamma & \cos\alpha\cos\gamma & \sin\alpha\sin\beta \\ -\sin\alpha\cos\beta\cos\gamma & -\sin\alpha\cos\beta\sin\gamma & \\ \sin\alpha\sin\gamma & -\sin\alpha\cos\gamma & \cos\alpha\sin\beta \\ -\cos\alpha\cos\beta\cos\gamma & -\cos\alpha\cos\beta\sin\gamma & \end{bmatrix} \tag{1.3}$$

where α is the knee flexion angle, γ is the tibial external rotation angle, and β is ($\pi/2\pm$ abduction); positive sign indicates a right knee and negative sign indicates a left knee.

Contact and Geometric Compatibility Conditions

As indicated in the introductory section of this chapter, several methods have been reported in the literature to provide three-dimensional mathematical representations of the articular surfaces of the femur and tibia.[12,21,68,124] In this model, and for simplicity, geometric primitives are employed. The coordinates of a sufficient number of points on the femoral condyles and tibial plateaus of several cadaveric knee specimens were obtained from related studies.[67,136] A separate mathematical function was determined as an approximate representation for each of the medial femoral condyle, the lateral femoral condyle, the medial tibial plateau, and the lateral tibial plateau. The femoral articular surfaces were approximated as parts of spheres, while the tibial plateaus were considered as planar surfaces as shown in Figs. 1.2 and 1.3. The equations of the medial and lateral femoral spheres expressed in the femoral coordinate system of axes were written as:

$$f(x, y) = -\sqrt{r^2 - (x - h)^2 - (y - k)^2} + 1 \tag{1.4}$$

where values of parameters (r, h, k, and l) were obtained as 21, 23.75, 18.0, 12.0 mm and 20.0, 23.0, 16.0, 11.5 mm for the medial and lateral spheres, respectively. The equations of the medial and lateral tibial planes expressed in the tibial coordinate system of axes were written as:

$$g(x', y') = my' + c \tag{1.5}$$

where values of parameters (m, c) were obtained as 0.358, 213 mm and –0.341, 212.9 mm for the medial and lateral planes, respectively.

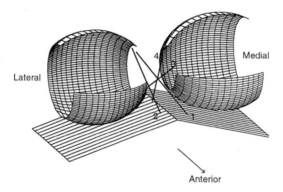

FIGURE 1.2   Three-dimensional model of the knee joint showing the anterior and posterior cruciate ligaments. (1) AAC, Anterior fibers of the anterior cruciate; (2) PAC, posterior fibers of the anterior cruciate; (3) APC, anterior fibers of the posterior cruciate; (4) PPC, posterior fibers of the posterior cruciate.

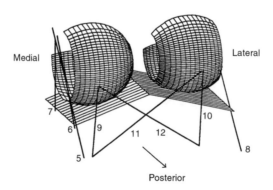

FIGURE 1.3   Three-dimensional model of the knee joint showing the collateral ligaments and the capsular structures. (5) AMC, anterior fibers of the medial collateral; (6) OMC, oblique fibers of the medial collateral; (7) DMC, deep fibers of the medial collateral; (8) LCL, lateral collateral; (9) MCAP, medial fibers of the posterior capsule; (10) LCAP, lateral fibers of the posterior capsule; (11) OPL, oblique popliteal ligament; (12) APL, arcuate popliteal ligament. (*Source*: Rahman, E.M. and Hefzy, M.S., *JJBE: Med. Eng. Physics*, 20, 4, 276, 1998. With permission from Elsevier Science.)

This model accomodates two situations: a two-point contact and a single-point contact. Initially, a two-point contact situation is assumed with the femur and tibia in contact on both medial and lateral sides. In the calculations, if one contact force becomes negative, then the two bones within its compartment are assumed to be separated, and the single-point contact situation is introduced, thus maintaining contact in the other compartment.

The contact condition requires that the position vectors of each contact point in the femoral and the tibial coordinate systems, $\vec{R}_c$ and $\vec{r}_c$, respectively, satisfy Eq. (1.2) as follows:

$$\vec{R}_c = \mathbf{R}_0 + [R]\vec{r}_c \tag{1.6}$$

where

$$\vec{R}_c = x_c\hat{i} + y_c\hat{j} + z_c\hat{k} \tag{1.6.a}$$

$$\vec{r}_c = x_c'\hat{i}' + y_c'\hat{j}' + z_c'\hat{k}' \tag{1.6.b}$$

where x_c, y_c, z_c and x_c', y_c', z_c' are the coordinates of the contact points in the femoral and tibial systems, respectively. Since contact occurs at points identifiable in both the femoral and tibial articulating surfaces, we can write at each contact point:

$$z_c = f(x_c, y_c) \tag{7.a}$$

$$z_c' = g(x_c', y_c') \tag{7.b}$$

where $f(x_c, y_c)$ and $g(x_c', y_c')$ are given by Eqs. (1.4) and (1.5), respectively. Eq. (1.6) can thus be rewritten as three scalar equations:

$$x_c = x_0 + R_{11}x_c' + R_{12}y_c' + R_{13}g(x_c', y_c') \tag{1.8.a}$$

$$y_c = y_0 + R_{21}x_c' + R_{22}y_c' + R_{23}g(x_c', y_c') \tag{1.8.b}$$

$$f(x_c, y_c) = z_0 + R_{31}x_c' + R_{32}y_c' + R_{33}g(x_c', y_c') \tag{1.8.c}$$

where R_{ij} is the *ijth* component of the rotational transformation matrix (R). Eqs. (1.8a) through (1.8c) constitute a mathematical definition for a contact point. Satisfying these equations at some given point will ensure that it is a contact point. Thus, in the two-point contact version of the model, Eqs. (1.8a) through (1.8c) generate six scalar equations which represent the contact conditions. In the one-point contact version of the model, Eqs. (1.8a) through (1.8c) produce three scalar equations which represent the contact conditions.

The geometric condition of compatibility of rigid bodies requires that a single tangent plane exists to both femoral and tibial surfaces at each contact point. This condition also implies that the normals to the femoral and tibial surfaces at each contact point are always colinear, and their cross product must vanish.

In order to express the geometric compatibility condition in a mathematical form, the position vector of the contact point in the femoral coordinate system (Eq. 1.6.a) is differentiated with respect to the local (x and y) coordinates to obtain two tangent vectors along these local directions. Cross product of these two tangent vectors is then employed to determine the unit vector normal to the femoral surface, $\hat{n}_f$, at the contact point. Using Eq. (1.7a), this unit vector is expressed in the femoral coordinate system as:

$$\hat{\mathbf{n}}_f = \frac{\dfrac{\partial f}{\partial x}\bigg|_{(x_c, y_c)}\hat{\mathbf{i}} + \dfrac{\partial f}{\partial y}\bigg|_{(x_c, y_c)}\hat{\mathbf{j}} - \hat{\mathbf{k}}}{\sqrt{1 + \left(\dfrac{\partial f}{\partial x}\right)^2 + \left(\dfrac{\partial f}{\partial y}\right)^2}} \qquad @ (x, y) = (x_c, y_c) \tag{1.9}$$

A similar analysis is performed to obtain the unit vector normal to the tibial surface, $\hat{\mathbf{n}}_t'$, at the contact point. Using Eq. (1.7b), this unit vector is expressed in the tibial coordinate system as:

$$\hat{\mathbf{n}}_t' = \frac{\dfrac{\partial g}{\partial x'}\bigg|_{(x_c', y_c')}\hat{\mathbf{i}}' + \dfrac{\partial g}{\partial y'}\bigg|_{(x_c', y_c')}\hat{\mathbf{j}}' - \hat{\mathbf{k}}'}{\sqrt{1 + \left(\dfrac{\partial g}{\partial x'}\right)^2 + \left(\dfrac{\partial g}{\partial y'}\right)^2}} \qquad @ (x', y') = (x_c', y_c') \tag{1.10}$$

Applying the rotational transformation matrix to Eq. (1.10) yields the unit normal vector to the tibial surface, $\hat{\mathbf{n}}_t$, expressed in the femoral coordinate system as:

$$\hat{\mathbf{n}}_t = \left(R_{11}n_{tx'}' + R_{12}n_{ty'}' - R_{13}n_{tz'}'\right)\hat{\mathbf{i}} + \left(R_{21}n_{tx'}' + R_{22}n_{ty'}' - R_{23}n_{tz'}'\right)\hat{\mathbf{j}}$$
$$+ \left(R_{31}n_{tx'}' + R_{32}n_{ty'}' - R_{33}n_{tz'}'\right)\hat{\mathbf{k}} \tag{1.11}$$

Since the unit vectors normal to the surfaces of the femur and tibia are colinear, they are equal:

$$\hat{\mathbf{n}}_f = \hat{\mathbf{n}}_t \tag{1.12}$$

This vectorial equation is rewritten in a scalar form as:

$$n_{fx} = \left(R_{11}n_{tx'}' + R_{12}n_{ty'}' - R_{13}n_{tz'}'\right) \quad @ \quad (x, y) = (x_c, y_c) \quad \text{and} \quad (x', y') = (x_c', y_c') \tag{1.13a}$$

$$n_{fy} = \left(R_{21}n_{tx'}' + R_{22}n_{ty'}' - R_{23}n_{tz'}'\right) \quad @ \quad (x, y) = (x_c, y_c) \quad \text{and} \quad (x', y') = (x_c', y_c') \tag{1.13b}$$

Eqs. (1.13a) and (1.13b) represent the geometric compatibility conditions at each contact point. Thus, for each contact point, two independent scalar equations are written generating four scalar equations to represent the geometric compatibility conditions in the two-point contact situation and two scalar equations to represent the geometric compatibility conditions in the one-point contact situation.

Ligamentous Forces

In this analysis, external loads are applied, and ligamentous and contact forces are then determined. The model includes 12 nonlinear spring elements that represent the different ligamentous structures and the capsular tissue posterior to the knee joint. Four elements represent the respective anterior and posterior fiber bundles of the anterior cruciate ligament (ACL) and the posterior cruciate ligament (PCL); three elements represent the anterior, deep, and oblique fiber bundles of the medial collateral ligament (MCL); one element represents the lateral collateral ligament (LCL); and four elements represent the medial, lateral, and oblique fiber bundles of the posterior part of the capsule (CAP). These twelve elements are shown in Figs. 1.2 and 1.3. The coordinates of the femoral and tibial insertion sites of the different ligamentous structures were specified according to the data available in the literature.[20,36] These coordinates are listed in Table 1.1.

TABLE 1.1   Local Attachment Coordinates of the Ligamentous Structures of the Present Model

Ligament	Femur			Tibia		
	x (mm)	y (mm)	z (mm)	x′ (mm)	y′ (mm)	z′ (mm)
ACL, ant. fibers	7.25	−15.6	21.25	−7.0	5.0	211.25
ACL, post. fibers	7.25	−20.3	19.55	2.0	2.0	212.25
PCL, ant. fibers	−4.75	−11.2	14.05	5.0	−30.0	206.25
PCL, post. fibers	−4.75	−23.2	15.65	−5.0	−30.0	206.25
MCL, ant. fibers	−34.75	−1.0	26.25	−20.0	4.0	171.25
MCL, oblique fibers	−34.75	−8.0	24.25	−35.0	−30.0	199.25
MCL, deep fibers	−34.75	−5.0	21.25	−35.0	0.0	199.25
LCL	35.25	−15.0	21.25	45.0	−25.0	176.25
Post. capsule, med.	−24.75	−38.0	6.25	−25.0	−25.0	181.25
Post. capsule, lat.	25.25	−35.5	8.25	25.0	−25.0	181.25
Post. capsule, oblique popliteal ligament	25.25	−35.5	8.25	−25.0	−25.0	181.25
Post. capsule, arcuate popliteal ligament	−24.75	−38.0	6.25	25.0	−25.0	181.25

ACL: Anterior Cruciate Ligament.
PCL: Posterior Cruciate Ligament.
MCL: Medial Collateral Ligament.
LCL: Lateral Collateral Ligament.
(*Source*: Rahman, E.M. and Hefzy, M.S., *JJBE: Med. Eng. Physics*, 20, 4, 276, 1998. With permission from Elsevier Science.)

In the present analysis, ligament wrapping around bone was not taken into consideration. The spring elements representing the ligamentous structures were thus assumed to be line elements extending from the femoral origin to the tibial insertion. These elements were assumed to carry load only when they are in tension, that is, when their length is larger than their slack, unstrained length, L_o. Ligaments exhibit a region of nonlinear force-elongation relationship, the "toe" region, in the initial stage of ligament strain, then a linear force-elongation relationship in later stages.[134] A two-piece force-elongation relationship was thus used to evaluate the magnitudes of the ligamentous forces.[21-23,118,129] This relationship is composed of two regions: a linear region and a parabolic region. The magnitude of the force in the *jth* ligamentous element is thus expressed as:

$$F_j = \begin{cases} 0 & ; \ \varepsilon_j \leq 0 \\ K1_j\left(L_j - L_{oj}\right)^2 & ; \ 0 \leq \varepsilon_j \leq 2\varepsilon_1 \\ K2_j\left(L_j - \left(1 + \varepsilon_1\right)L_{oj}\right)^2 & ; \ \varepsilon_j \geq 2\varepsilon_1 \end{cases} \tag{1.14}$$

where $K1_j$ and $K2_j$ are the stiffness coefficients of the *jth* spring element for the parabolic and linear regions, respectively, and L_j and L_{oj} are its current and slack lengths, respectively. The strain in the *jth* ligamentous element, ε_j, is given by

$$\varepsilon_j = \frac{L_j - L_{oj}}{L_{oj}} \tag{1.15}$$

and the linear range threshold is specified as $\varepsilon_1 = 0.03$.

Values of the stiffness coefficients of the spring elements used to model the different ligamentous structures were estimated according to the data available in the literature[21,23,30,93-96,109,118,129,130,133] and are listed in Table 1.2. The slack length of each spring element is obtained by assuming an extension ratio e_j at full extension and using the following relation:

TABLE 1.2   Stiffness Coefficients of the Spring Elements Representing the
Ligamentous Structures of the Present Model

Ligament	K1 (N/mm²)	K2 (N/mm²)
ACL, ant. fibers	83.15	22.48
ACL, post. fibers	83.15	26.27
PCL, ant. fibers	125.00	31.26
PCL, post. fibers	60.00	19.29
MCL, ant. fibers	91.25	10.00
MCL, oblique fibers	27.86	5.00
MCL, deep fibers	21.07	5.00
LCL	72.22	10.00
Post. capsule, med.	52.59	12.00
Post. capsule, lat.	54.62	12.00
Post. Capsule, oblique popliteal ligament	21.42	3.00
Post. Capsule, arcuate popliteal ligament	20.82	3.00

ACL: Anterior Cruciate Ligament.
PCL: Posterior Cruciate Ligament.
MCL: Medial Collateral Ligament.
LCL: Lateral Collateral Ligament.
(*Source*: Rahman, E.M. and Hefzy, M.S., *JJBE: Med. Eng. Physics*, 20, 4, 276,
1998. With permission from Elsevier Science.)

$$\varepsilon_j = \frac{L_j \big|_{\text{at full extension}}}{L_{oj}} \tag{1.16}$$

to evaluate the spring element's slack length, L_{oj}, from its length at full extension (which can be calculated using the coordinates of the attachment points). The values of the extension ratios were specified according to the data available in the literature[20,60] and are listed in Table 1.3.

TABLE 1.3   Extension Ratios at Full Extension of the
Ligamentous Structures of the Present Model

Ligament	e
ACL, ant. fibers	1.000
ACL, post. fibers	1.051
PCL, ant. fibers	1.004
PCL, post. fibers	1.050
MCL, ant. fibers	0.940
MCL, oblique fibers	1.031
MCL, deep fibers	1.049
LCL	1.050
Post. capsule, med.	1.080
Post. capsule, lat.	1.080
Post. capsule, oblique popliteal ligament	1.080
Post. capsule, arcuate popliteal ligament	1.070

ACL: Anterior Cruciate Ligament.
PCL: Posterior Cruciate Ligament.
MCL: Medial Collateral Ligament.
LCL: Lateral Collateral Ligament.
(*Source*: Rahman, E.M. and Hefzy, M.S., *JJBE: Med. Eng.
Physics*, 20, 4, 276, 1998. With permission from Elsevier
Science.)

Contact Forces

As the tibia moves with respect to the femur, the contact points also move in the respective medial and lateral compartments. Contact forces are induced at one or both contact points. These forces are applied normal to the articular surface. Thus, the contact force applied to the tibia is expressed as: $\vec{N}_i = N_i \hat{n}_i$ where N_i is the magnitude of the contact force, and $\hat{n}_i$ is the unit vector normal to the tibial surface at the contact point, expressed in the femoral coordinate system. In the two-point contact situation, $i = 1, 2$ and in the single-point contact situation, $i = 1$.

Equations of Motion

The equations governing the three-dimensional motion of the tibia with respect to the femur are the second order differential Newton's and Euler's equations of motion. Newton's equations are written in a scalar form, with respect to the femoral fixed system of axes, as:

$$F_{ex} + W_x + \sum_{i=1}^{2} N_{ix} + \sum_{j=1}^{12} F_{jx} = m\,\ddot{x}_o \tag{1.17}$$

$$F_{ey} + W_y + \sum_{i=1}^{2} N_{iy} + \sum_{j=1}^{12} F_{jy} = m\,\ddot{y}_o \tag{1.18}$$

$$F_{ez} + W_z + \sum_{i=1}^{2} N_{iz} + \sum_{j=1}^{12} F_{jz} = m\,\ddot{z}_o \tag{1.19}$$

where m is the mass of the leg, $\ddot{x}_o$, $\ddot{y}_o$, and $\ddot{z}_o$ are the components of the linear acceleration of the center of mass of the leg (in the fixed femoral coordinate system); W_x, W_y, and W_z are the components of the weight of the leg; and F_{ex}, F_{ey}, and F_{ez} are the components of the external forcing pulse applied to the tibia.

Euler's equations of motion are written with respect to the moving tibial system of axes which is the tibial centroidal principal system of axes (x′, y′ and z′). Thus, the angular velocity components ($\dot{\theta}_{x'}$, $\dot{\theta}_{y'}$, $\dot{\theta}_{z'}$) and angular acceleration components ($\ddot{\theta}_{x'}$, $\ddot{\theta}_{y'}$, $\ddot{\theta}_{z'}$), in the Euler equations, are expressed with respect to this principal system of axes as:

$$\dot{\theta}_{x'} = -\dot{\alpha}\sin\beta\cos\gamma - \dot{\alpha}\beta\cos\beta\cos\gamma + \dot{\alpha}\gamma\sin\beta\sin\gamma + \dot{\beta}\sin\gamma + \dot{\beta}\gamma\cos\gamma \tag{1.20a}$$

$$\dot{\theta}_{y'} = -\dot{\alpha}\sin\beta\sin\gamma - \dot{\alpha}\beta\cos\beta\sin\gamma - \dot{\alpha}\gamma\sin\beta\cos\gamma + \dot{\beta}\cos\gamma + \dot{\beta}\gamma\sin\gamma \tag{1.20b}$$

$$\dot{\theta}_{z'} = -\dot{\alpha}\cos\beta + \dot{\alpha}\beta\sin\beta - \dot{y} \tag{1.20c}$$

$$\ddot{\theta}_{x'} = -\ddot{\alpha}\sin\beta\cos\gamma - \ddot{\alpha}\beta\cos\beta\cos\gamma + \ddot{\alpha}\gamma\sin\beta\sin\gamma - \dot{\alpha}^2\beta\sin\gamma - \dot{\alpha}^2\gamma\sin\beta\cos\beta\cos\gamma$$
$$-2\dot{\alpha}\beta\cos\beta\cos\gamma + 2\dot{\alpha}\beta\gamma\cos\beta\sin\gamma + 2\dot{\alpha}\gamma\sin\beta\sin + \beta\sin\gamma \tag{1.21a}$$
$$+\ddot{\beta}\gamma\cos\gamma + 2\ddot{\beta}\dot{\gamma}\cos\gamma$$

$$\ddot{\theta}_{y'} = -\ddot{\alpha}\sin\beta\sin\gamma - \ddot{\alpha}\beta\cos\beta\sin\gamma - \ddot{\alpha}\gamma\sin\beta\cos\gamma - \dot{\alpha}^2\beta\cos\gamma - \dot{\alpha}^2\gamma\sin\beta\cos\beta\cos\gamma$$
$$-2\dot{\alpha}\beta\cos\beta\cos\gamma + 2\dot{\alpha}\dot{\beta}\gamma\cos\beta\cos\gamma - 2\dot{\alpha}\dot{\gamma}\sin\beta\cos\gamma - \ddot{\beta}\sin\gamma \tag{1.21b}$$
$$+\ddot{\beta}\gamma\sin\gamma + 2\ddot{\beta}\dot{\gamma}\sin\gamma$$

$$\ddot{\theta}_{z'} = \ddot{\alpha}\cos\beta + \ddot{\alpha}\beta\sin\beta + \dot{\alpha}^2\gamma\sin^2\beta + 2\dot{\alpha}\beta\sin\beta + \dot{\beta}^2\gamma - \ddot{y} \tag{1.21c}$$

Euler's equations are written in a scalar form as:

$$(\Sigma M)_{x'} = I_{x'x'}\ddot{\theta}_{x'} + \left(I_{z'z'} - I_{y'y'}\right)\dot{\theta}_{y'}\dot{\theta}_{z'} \tag{1.22}$$

$$(\Sigma M)_{y'} = I_{y'y'}\ddot{\theta}_{y'} + \left(I_{x'x'} - I_{z'z'}\right)\dot{\theta}_{x'}\dot{\theta}_{z'} \tag{1.23}$$

$$(\Sigma M)_{z'} = I_{z'z'}\ddot{\theta}_{z'} + \left(I_{y'y'} - I_{x'x'}\right)\dot{\theta}_{x'}\dot{\theta}_{y'} \tag{1.24}$$

where $(\Sigma M)_{x'}$, $(\Sigma M)_{y'}$, and $(\Sigma M)_{z'}$ are the sum of the moments of all forces acting on the tibia around the x', y', and z' axis, respectively, and $I_{x'x'}$, $I_{y'y'}$, and $I_{z'z'}$ are the principal moments of inertia of the leg about this centroidal principal axis. The inertial parameters were estimated using the anthropometric data available in the literature.[32,34,40,102,120] In this analysis, the mass of the leg was taken as m = 4.0 kg. Also, the leg was assumed to be a right cylinder; mass moments of inertia were thus calculated as $I_{x'x'}$ = 0.0672 kg m^2, $I_{y'y'}$ = 0.0672 kg m^2, and $I_{z'z'}$ = 0.005334 kg m^2.

In the two point-contact situation, tibio-femoral motions are thus described in terms of six differential equations of motion: Eqs. (1.17) through (1.19) and (1.22) through (1.24), and ten nonlinear algebraic equations [six contact conditions: Eqs. (1.8a through 1.8c), and four geometric compatibility conditions: Eqs. (1.13a) and (1.13b)]. In the one-point contact situation, the ten algebraic equations reduce to five: three contact conditions and two geometric compatibility conditions. The governing system of equations in the two-point contact version of the model thus consists of 16 equations in 16 unknowns: six motion parameters (x_o, y_o, z_o, α, β, and γ); two contact forces (N_1 and N_2); and eight contact parameters [(x_{c1}, y_{c1}) and (x_{c2}, y_{c2}): the coordinates of the medial and lateral contact points in the femoral system of axes, respectively, and (x_{c1}', y_{c1}') and (x_{c2}', y_{c2}'): the coordinates of the medial and lateral contact points in the tibial system of axes, respectively]. In the one-point contact version of the model, the governing system of equations reduces to 11 equations in 11 unknowns.

1.3   Solution Algorithm

In this formulation, six second order ordinary differential equations (ODEs), Newton's and Euler's equations of motion, are written to describe the general motion of the tibia with respect to the femur. Two-point contact is initially assumed. At each contact point five nonlinear algebraic constraints are written to satisfy the contact and compatibility conditions. Thus, this system of equations can be expressed as:

$$\vec{F}(\ddot{\vec{y}},\ \dot{\vec{y}},\ \vec{y}, t) \tag{1.25}$$

where $\dot{\vec{y}} = \dfrac{d\vec{y}}{dt}$ and $\ddot{\vec{y}} = \dfrac{d\dot{\vec{y}}}{dt}$. This system has two parts: a differential part and an algebraic part. These ODE systems are called differential-algebraic equations (DAEs). Numerical methods from the field of ODEs have classically been employed to solve DAE systems.[24,53-56,105]

The behavior of the numerical methods used to solve a DAE system depends on the DAE's index. While the existing DAE algorithms are robust enough to handle systems of index one, they encounter difficulties in solving systems of higher indices. The index of a DAE system is the number of times the algebraic constraints need to be differentiated in order to match the order of the differential part of the system and at the same time be able to solve the DAE system for explicit expressions for each of the components of the vector $\ddot{\vec{y}}$.[55] In the present system, N_1 and N_2, two independent variables in vector $\vec{y}$, appear only in the differential equations of motion. In order to generate terms that include $\dot{N}_1$ and $\dot{N}_2$, which are components of vector $\dot{\vec{y}}$, the differential equations need to be differentiated once more with respect to time. These equations are then transformed to third order differential equations. The algebraic constraints are then differentiated thrice with respect to time in order to match the order of the differential part of the system. Consequently, the present system of equations describing the dynamic behavior of the knee joint has an index value of three.

To reduce the order of the system it is rewritten in an equivalent mathematical form which has the same analytical solution but possesses a lower index. The second order time derivatives, accelerations in the equations of motion, are transformed to first order time derivatives of velocity. The system is then augmented with six more first order ordinary differential equations relating the joint motions to the joint velocities. The differential part of the system is transformed to

$$v_x = \dot{x}_o \tag{1.26a}$$

$$v_y = \dot{y}_o \tag{1.26b}$$

$$v_z = \dot{z}_o \tag{1.26c}$$

$$\omega_\alpha = \dot{\alpha} \tag{1.26d}$$

$$\omega_\beta = \dot{\beta} \tag{1.26e}$$

$$\omega_\gamma = \dot{\gamma} \tag{1.26f}$$

$$F_{ex} + W_x + \sum_{i=1}^{2} N_{ix} + \sum_{j=1}^{12} F_{jx} = m\,\dot{v}_x \tag{1.27a}$$

$$F_{ey} + W_y + \sum_{i=1}^{2} N_{iy} + \sum_{j=1}^{12} F_{jy} = m\,\dot{v}_y \tag{1.27b}$$

$$F_{ez} + W_z + \sum_{i=1}^{2} N_{iz} + \sum_{j=1}^{12} F_{jz} = m\,\dot{v}_z \tag{1.27c}$$

$$(\Sigma M)_{x'} = I_{xx'}\dot{\omega}_{x'} + \left(I_{zz'} - I_{yy'}\right)\omega_{y'}\omega_{z'} \tag{1.27d}$$

$$(\Sigma M)_{y'} = I_{yy'}\dot{\omega}_{y'} + \left(I_{xx'} - I_{zz'}\right)\omega_{x'}\omega_z \tag{1.27e}$$

$$(\Sigma M)_{z'} = I_{zz'}\dot{\omega}_{z'} + \left(I_{yy'} - I_{xx'}\right)\omega_{x'}\omega_y \tag{1.27f}$$

Using Eqs. (1.26d) through (1.26f) into Eqs. (1.20) and (1.21), $\omega_{x'}$, $\omega_{x'}$, $\omega_{x'}$, $\dot{\omega}_{x'}$, $\dot{\omega}_{x'}$, and $\dot{\omega}_{x'}$ are expressed as:

$$\omega_{x'} = -\omega_\alpha \sin\beta \cos\gamma - \omega_\alpha \beta \cos\beta \cos\gamma + \omega_\alpha \gamma \sin\beta \sin\gamma + \omega_\beta \sin\gamma + \omega_\beta \gamma \cos\gamma \tag{1.28a}$$

$$\omega_{y'} = -\omega_\alpha \sin\beta \sin\gamma - \omega_\alpha \beta \cos\beta \sin\gamma - \omega_\alpha \gamma \sin\beta \cos\gamma - \omega_\beta \cos\gamma + \omega_\beta \gamma \sin\gamma \tag{1.28b}$$

$$\omega_{z'} = -\omega_\alpha \cos\beta + \omega_\alpha \beta \sin\beta - \omega_\gamma \qquad (1.28c)$$

$$\dot{\omega}_{x'} = -\dot{\omega}_\alpha \sin\beta \cos\gamma - \dot{\omega}_\alpha \beta \cos\beta \cos\gamma + \dot{\omega}_\alpha \gamma \sin\beta \sin\gamma - \omega_\alpha^2 \beta \sin\gamma - \omega_\alpha^2 \gamma \sin\beta \cos\beta \cos\gamma$$

$$-2\omega_\alpha \omega_\beta \cos\beta \cos\gamma + 2\omega_\alpha \omega_\beta \gamma \cos\beta \sin\gamma + 2\omega_\alpha \omega_\gamma \sin\beta \sin\gamma + \dot{\omega}_\beta \sin\gamma$$

$$+ \dot{\omega}_\beta \gamma \cos\gamma + 2\omega_\beta \omega_\gamma \cos\gamma$$

$$(1.29a)$$

$$\dot{\omega}_{y'} = -\dot{\omega}_\alpha \sin\beta \sin\gamma - \dot{\omega}_\alpha \beta \cos\beta \sin\gamma - \dot{\omega}_\alpha \gamma \sin\beta \cos\gamma + \omega_\alpha^2 \beta \cos\gamma - \omega_\alpha^2 \gamma \sin\beta \cos\beta \sin\gamma$$

$$-2\omega_\alpha \omega_\beta \cos\beta \sin\gamma + 2\omega_\alpha \omega_\beta \gamma \cos\beta \cos\gamma - 2\omega_\alpha \omega_\gamma \sin\beta \cos\gamma - \dot{\omega}_\beta \cos\gamma$$

$$+ \dot{\omega}_\beta \gamma \sin\gamma + 2\omega_\beta \omega_\gamma \sin\gamma$$

$$(1.29b)$$

$$\dot{\omega}_{z'} = \dot{\omega}_\alpha \cos\beta + \dot{\omega}_\alpha \beta \sin\beta + \omega_\alpha^2 \gamma \sin^2\beta + 2\omega_\alpha \omega_\beta \sin\beta + \omega_\alpha^2 \gamma - \dot{\omega}_\gamma \qquad (1.29c)$$

The resulting system of equations contains twelve first order ordinary differential equations and ten nonlinear algebraic constraints. It can be written as:

$$\vec{F}(\vec{y}, \dot{\vec{y}}, t) = \vec{0} \qquad (1.30)$$

where $\vec{y}$ is a vector of dimension ($n = 22$) containing the 22 independent variables, namely, $\{x_o, y_o, z_o, \alpha, \beta, \gamma, v_x, v_y, v_z, \omega_\alpha, \omega_\beta, \omega_\gamma, x_{c1}, y_{c1}, x_{c2}, y_{c2}, x_{c1}', y_{c1}', x_{c2}', y_{c2}', N_1,$ and $N_2\}$. This is a DAE system of 22 equations to be solved in 22 unknowns. While it is mathematically equivalent to the original system, it has an index of two. To generate $\dot{N}_1$ and $\dot{N}_2$, the equations of motion (which are now of order one) are to be differentiated once more bringing them to order two. The algebraic constraints are then differentiated twice with respect to time so they can match the order of the differential part of the system. Therefore, the new system of equations describing the dynamic behavior of the knee joint has an index value of two.

To solve the DAE system, the algorithm divides the analysis time span into time steps of variable size, h_i, and time stations, t_i. From time station t_n, the algorithm takes a step forward on time of size h_{n+1} to evaluate and $\vec{y}$ and $\dot{\vec{y}}$ at time station t_{n+1}; that is,

$$t_{n+1} = t_n + h_{n+1} \qquad (1.31)$$

At each time station t_{n+1}, components of $\dot{\vec{r}}_{n+1}$ are approximated in terms of $\vec{y}_{n+1}$ and $\vec{y}$ at previous time steps using an integration formula such as a backward differentiation formula (BDF) or a Runge-Kutta (R-K) method. The most popular integration scheme is multistep BDF. This scheme yields:

$$\dot{\vec{y}}_{n+1} = \frac{1}{h_{n+1}} \sum_{i=0}^{k} \alpha_i \vec{y}_{n+1-i} \qquad (1.32)$$

where k is the order of the formula and (α_i, $i = 0, k$) are the coefficients of the BDF. These coefficients are determined using generating polynomials such as those defined by Jackson and Sacks-Davis.[77] The order of the BDF formula is equal to the number of steps over which it extrapolates the solution. At every step h_{n+1}, the order, k, and step size, h, used are dependent on the behavior of the solution.

Eq. (1.32) is hence used to eliminate $\dot{\vec{y}}$ from the system of equations, and the DAE system defined in Eq. (1.30) is transformed to the nonlinear algebraic system:

$$\vec{F}\left(\vec{y}_{n+1}, t_{n+1}\right) = \vec{0} \tag{1.33}$$

A variation of the Newton-Raphson iteration technique is then used to solve the nonlinear system for $\vec{y}_{n+1}$.[80] A solution for the resulting system is thus obtained using the differential form of the Newton-Raphson method which includes evaluating iteratively $\{\Delta\vec{y}_{(i)}\}$ by solving the following system of equations:

$$\left[K\left(\vec{y}, t\right)\right]\left\{\Delta\vec{y}_{(1)}\right\} = \left\{\vec{R}\left(\vec{y}, t\right)\right\} \tag{1.34}$$

where

$$\left[K\left(\vec{y}, t\right)\right] = \left[\frac{\partial\vec{F}}{\partial\vec{y}}\bigg|_{\left(\vec{y}_{n+1,0}, t_{n+1}\right)}\right] + \frac{\alpha_s}{h_{n+1}}\left[\frac{\partial\vec{F}}{\partial\vec{y}}\bigg|_{\left(\vec{y}_{n+1,0}, t_{n+1}\right)}\right] \tag{1.35a}$$

and

$$\left\{\vec{R}\left(\vec{y}, t\right)\right\} = -\left\{\vec{F}\left(\vec{y}_{n+1,(i-1)}, \dot{\vec{y}}_{n+1,(i-1)}, t_{n+1}\right)\right\} \tag{1.35b}$$

where $\vec{F}$ is defined by Eq. (1.30). After each iteration $\vec{y}$ is updated according to:

$$\vec{y}_{n+1,(i)} = \vec{y}_{n+1,(i-1)} + \Delta\vec{y}_{(i)} \tag{1.36}$$

The iterations continue until $\Delta\vec{y}_{(i)}$ satisfies a pre-set convergence criterion. A local error test is then carried out to check whether the solution satisfies user-defined error parameters. If the solution is acceptable, the converged values of $\vec{y}_{N+1}$ and $\dot{\vec{y}}_{n+1}$ are then used with values of and $\vec{y}$ and $\dot{\vec{y}}$ at the previous k time stations to evaluate $\vec{y}$ and $\dot{\vec{y}}$ at t_{n+2} and the algorithm continues marching on in time. The stiffness matrix (K) used in step h_{n+1} is carried on unchanged to step h_{n+2} unless the algorithm fails to complete step h_{n+1} successfully. If the Newton-Raphson iterations fail to converge or the converged solution fails to satisfy the user-defined error parameters, the algorithm goes back to t_n and retakes the step with an updated stiffness matrix, a smaller step size h, and/or a BDF of a different order.

The initial guess ($i = 0$) for $\vec{y}_{N+1}$ and $\dot{\vec{y}}_{n+1}$ (required to begin the Newton-Raphson iterations) is predicted based on values of $\vec{y}$ at the previous $k+1$ time stations for a kth order integration formula. A predictor polynomial, which interpolates $\vec{y}$ at the previous $k+1$ time stations, is used to extrapolate the values of $\vec{y}$ at time station t_{n+1} while its derivative is used to extrapolate the values of $\dot{\vec{y}}$ at time station t_{n+1}. The extrapolation polynomial[24] can be written as:

$$\vec{w}_{n+1}(t) = \vec{y}_n + \left(t - t_n\right)\left[\vec{y}_n, \vec{y}_{n-1}\right] + \left(t - t_n\right)\left(t - t_{n-1}\right)\left[\vec{y}_n, \vec{y}_{n-1}, \vec{y}_{n-2}\right]$$
$$+ \ldots + \left(t - t_n\right)\left(t - t_{n-1}\right)\ldots\left(t - t_{n-k+1}\right)\left[\vec{y}_n, \vec{y}_{n-1}, \ldots, \vec{y}_{n-k}\right] \tag{1.37}$$

where the recurrence formula of the divided differences is given by

$$\left.\begin{array}{l}\left[\vec{\mathbf{y}}_{n}\right]=\vec{\mathbf{y}}_{n}\\[2em]\left[\vec{\mathbf{y}}_{n},\vec{\mathbf{y}}_{n-1},\ldots,\vec{\mathbf{y}}_{n-k}\right]=\dfrac{\left[\vec{\mathbf{y}}_{n},\vec{\mathbf{y}}_{n-1},\ldots,\vec{\mathbf{y}}_{n-k+1}\right]-\left[\vec{\mathbf{y}}_{n-1},\vec{\mathbf{y}}_{n-2},\ldots,\vec{\mathbf{y}}_{n-k}\right]}{t_n-t_{n-k}}\end{array}\right\} \qquad (1.38)$$

Using Eqs. (1.37) and (1.38), $\vec{\mathbf{y}}_{n+1}$ and $\dot{\vec{\mathbf{y}}}_{n+1}$ are estimated by evaluating $\vec{\mathbf{w}}_{n+1}(t)$ and $\dot{\vec{\mathbf{w}}}_{n+1}(t)$, respectively, at $(t = t_{n+1})$. This is called the predictor part of the algorithm while the solution of the nonlinear algebraic system through Newton-Raphson iterations is called the corrector part of the algorithm. Algorithms which use this approach are called predictor-corrector algorithms.

The initial values of $\vec{\mathbf{y}}$ and $\dot{\vec{\mathbf{y}}}$ at $t = t_0$ (namely, $\vec{\mathbf{y}}_0$ and $\dot{\vec{\mathbf{y}}}_0$) must be specified in order to completely define this initial value problem. These initial conditions must be consistent, i.e., they must satisfy the DAE system at time $t = t_0$:

$$\vec{\mathbf{F}}\left(\dot{\vec{\mathbf{y}}}_0, \vec{\mathbf{y}}_0, t_0\right) = \vec{\mathbf{0}} \qquad (1.39)$$

It is important to realize that DAE solvers are very sensitive to the initial conditions. A small inconsistency in the initial conditions, especially for an index two DAE system, can cause the algorithm to diverge in the first step.[24]

DAE Solvers

Two major computer codes have been developed by the Lawrence Livermore National Laboratory to integrate a DAE using the BDF: the Differential/Algebraic System Solver (DASSL) developed by Petzold[106] and the Livermore Solver for Ordinary Differential Equations: Implicit form (LSODI) developed by Hindmarsh and Painter.[71] Both codes use error-controlled, variable-step, variable-order (integration formula) predictor-corrector algorithms to integrate the DAE. Starting at time station (t_n), the predictor extrapolates the values of $\vec{\mathbf{y}}$ and $\dot{\vec{\mathbf{y}}}$ at time station (t_{n+1}) based on their values at earlier time stations using a forward differentiation formula. Then the corrector utilizes a BDF of order ranging from one to five to transform $\vec{\mathbf{F}}(\dot{\vec{\mathbf{y}}}_{n+1}, \vec{\mathbf{y}}_{n+1}, t_{n+1}) = \vec{\mathbf{0}}$ to the form $\vec{\mathbf{F}}(\vec{\mathbf{y}}_{n+1}, t_{n+1}) = \vec{\mathbf{0}}$. The two codes differ in the BDF formulas they use and in the step size, order selection, and error control strategies.[24] Both the load vector and the stiffness matrix used in LSODI and DASSL are similar. In both codes, a solution for the resulting system is then obtained using the differential form of the Newton-Raphson method which includes solving Eq. (1.34) iteratively. After each corrector iteration a convergence test is carried; and after convergence, a local error test is also carried.

We have used both LSODI and DASSL to obtain a solution for the present DAE system.[5-7] Some instabilities were encountered when using the LSODI which is consistent with reports that DASSL is more stable and robust.[24] Aside from having different error control strategies, the main difference between the two codes is that while the LSODI uses a fixed coefficient implementation of the BDF formulas [Eq. (1.32)], the DASSL uses a fixed leading coefficient implementation.[24] These two techniques allow the multistep-fixed step size BDF formulas to accomodate multistep-variable step size applications. In the fixed coefficient implementation, all coefficients of $\vec{\mathbf{y}}_{n+1-i}$ are unchanged, even when different step sizes, h_i, are used. In the fixed leading coefficient implementation, only the first coefficient [that of $\vec{\mathbf{y}}_{n+1}$ in Eq. (1.32)] does not depend on the step size. We like to point out that Hindmarsh, one of the authors of LSODI, indicated that the LSODI is essentially a stiff differential equation solver, and its use as a DAE solver is only marginal (personal communication). In what follows, we briefly introduce the DASSL software to the reader.

The DASSL computer code is a general purpose DAE solver designed to solve systems of indices zero and one. It can also solve some classes of higher index DAEs including semi-explicit index two systems such as the present DAE system.

Input to the DASSL includes the initial time where the integration begins $t = t_0$, the end of the integration interval $t = t_F$, $\vec{y}_0$, $\dot{\vec{y}}_0$, a relative error tolerance vector $\vec{\varepsilon}_{rel}$ and an absolute error tolerance vector $\vec{\varepsilon}_{abs}$. Each independent variable has a corresponding component in $\vec{\varepsilon}_{rel}$ and $\vec{\varepsilon}_{abs}$. User supplied subroutines evaluate the load vector $\vec{F}(\vec{y}, \dot{\vec{y}}, t)$ and the stiffness matrix $[\mathbf{K}(\vec{y}, t)]$. DASSL is called recurrently in a loop which updates t_F until the analysis time span is covered.

The integration formula used by DASSL is a variable step size h variable order k fixed leading coefficient α_s version of the BDF. The order of the BDF can range from one to five.[24] At the first time station t_0, the code estimates the initial time step size h_1 in terms of t_F, t_0, $\vec{y}_0$, $\dot{\vec{y}}_0$, $\vec{\varepsilon}_{rel}$, and $\vec{\varepsilon}_{abs}$. At each time station t_n, the predictor formula is used to evaluate $\vec{y}_{n+1,(0)}$, then the corrector iterations are used to correct this value. After each corrector iteration, a convergence test is carried out insuring that the weighted root mean square norm of $\Delta\vec{y}_{(i)}$ is less than a pre-set convergence constant. The default norm used in DASSL is a weighted root mean square norm, where the weights depend on the relative and absolute error tolerance vectors and on the value of $\vec{y}$ at the beginning of the step. If the convergence test is not satisfied after four iterations, the step is aborted. The algorithm goes back to station t_n, calculates the stiffness matrix, if it was not current, and repeats the step again. If it fails to converge again, the step size is reduced by a factor of one quarter. If, after ten consecutive step size reductions, the code still fails to take the step, or if at any time h becomes less than the minimum step size, h_{min}, the code is aborted with a fatal error. The minimum step size is either specified by the user or calculated by the code in terms of t_n, t_F, and the computer roundoff error. If the code is aborted with a fatal error, the user needs to modify the absolute and/or relative error vectors (error tolerance) and restart from the beginning.

If the convergence test is successful, a local error test is carried on the converged solution $\vec{y}_{n+1}$. The test amounts to requiring that the weighted root mean square norm of the difference between the converged solution, $\vec{y}_{n+1}$, and the predicted solution, $\vec{y}_{n+1,(0)}$, be less than the user-defined error tolerances. If the test fails, the step is aborted and the algorithm goes back to t_n and takes the step again.

Regardless of the success of the convergence test (or the local error test), an appropriate order of the BDF and a new step size are calculated before a new step is taken (or before the same step is retaken). The very first time step has to be taken with a BDF of order one (i.e., Euler's method). Thereafter, and for an initial number of subsequent steps (an initial phase), the code raises the order of the BDF, k, and increases the step size, h. After that initial period, it begins to estimate the errors for different orders by calculating the dominant terms in the remainder of a Taylor series expansion of the solution of order $k - 2$, $k - 1$, k and $k + 1$, respectively. If the weighted root mean square norm of these quantities (error estimates) decreases with the increase of the order, the order of the BDF, k, is increased by one and if the weighted root mean square norm of the estimates increases with the increase of the order, the order of the BDF, k, is decreased by one. A new step size is then calculated according to the estimate of the local error in the last successful step. The smaller the local error estimate, the larger the next step size will be (in comparison to the previous step). After a successful step, the change in the next step size never exceeds a maximum of double or a minimum of half the previous step size. After an unsuccessful attempted step, which as a result is retaken, the step size is decreased according to the local error estimate in the last successful step. If more than one unsuccessful step has been attempted successively, then the step size is decreased to 25% of its value.

After the local error test fails three times consecutively, the order of the BDF is reduced to one, since a BDF of order one is the most stable fixed leading coefficient BDF at small step size. While trying to satisfy the local error test, if h becomes smaller than h_{min} the code is aborted with a fatal error; the user must modify the error tolerance and restart from the beginning.

Load Vector and Stiffness Matrix

Expressions for the load vector, assuming contact on both the medial and lateral compartments, are determined from Eq. (1.35b) as:

$$F_1 = v_x - \dot{x}_o \tag{1.40}$$

$$F_2 = v_y - \dot{y}_o \tag{1.41}$$

$$F_3 = v_z - \dot{z}_o \tag{1.42}$$

$$F_4 = \omega_\alpha - \dot{\alpha} \tag{1.43}$$

$$F_5 = \omega_\beta - \dot{\beta} \tag{1.44}$$

$$F_6 = \omega_\gamma - \dot{\gamma} \tag{1.45}$$

$$F_7 = F_{ex} + W_x + \sum_{i=1}^{2} N_{ix} + \sum_{j=1}^{12} F_{jx} = m\dot{v}_x \tag{1.46}$$

$$F_8 = F_{ey} + W_y + \sum_{i=1}^{2} N_{iy} + \sum_{j=1}^{12} F_{jy} = m\dot{v}_y \tag{1.47}$$

$$F_9 = F_{ez} + W_z + \sum_{i=1}^{2} N_{iz} + \sum_{j=1}^{12} F_{jz} = m\dot{v}_z \tag{1.48}$$

$$F_{10} = (\Sigma M)_{x'} - I_{xx'}\dot{\omega}_{x'} - \left(I_{zz'} - I_{yy'}\right)\omega_{y'}\omega_{z'} \tag{1.49}$$

$$F_{11} = (\Sigma M)_{y'} - I_{yy'}\dot{\omega}_{y'} - \left(I_{xx'} - I_{zz'}\right)\omega_{x'}\omega_{z'} \tag{1.50}$$

$$F_{12} = (\Sigma M)_{z'} - I_{zz'}\dot{\omega}_{z'} - \left(I_{yy'} - I_{xx'}\right)\omega_{x'}\omega_{y} \tag{1.51}$$

$$F_{13} = -x_{c1} + x_o + R_{11}x'_{c1} + R_{12}y'_{c1} + R_{13}g_1\left(x'_{c1}, y'_{c1}\right) \tag{1.52}$$

$$F_{14} = -y_{c1} + y_o + R_{21}x'_{c1} + R_{22}y'_{c1} + R_{23}g_1\left(x'_{c1}, y'_{c1}\right) \tag{1.53}$$

$$F_{15} = -f_1\left(x_{c1}, y_{c1}\right) + z_o + R_{31}x'_{c1} + R_{32}y'_{c1} + R_{33}g_1\left(x'_{c1}, y'_{c1}\right) \tag{1.54}$$

$$F_{16} = -x_{c2} + x_o + R_{11}x'_{c2} + R_{12}y'_{c2} + R_{13}g_2\left(x'_{c2}, y'_{c2}\right) \tag{1.55}$$

$$F_{17} = -y_{c2} + y_o + R_{21}x'_{c2} + R_{22}y'_{c2} + R_{23}g_2\left(x'_{c2}, y'_{c2}\right) \tag{1.56}$$

$$F_{18} = -f_2\left(x_{c2}, y_{c2}\right) + z_o + R_{31}x'_{c2} + R_{32}y'_{c2} + R_{33}g_2\left(x'_{c2}, y'_{c2}\right) \tag{1.57}$$

$$F_{19} = \left(n_{fx}\right)_1 - \left[R_{11}\left(n'_{tx'}\right)_1 + R_{12}\left(n'_{ty'}\right)_1 - R_{13}\left(n'_{tz'}\right)_1\right]$$

$$@ \left(x, y\right) = \left(x_{c1}, y_{c1}\right) \quad \text{and} \quad \left(x', y'\right) = \left(x'_{c1}, y'_{c1}\right)$$

(1.58)

$$F_{20} = \left(n_{fy}\right)_1 - \left[R_{21}\left(n'_{tx'}\right)_1 + R_{22}\left(n'_{ty'}\right)_1 - R_{23}\left(n'_{tz'}\right)_1\right]$$

$$@ \left(x, y\right) = \left(x_{c1}, y_{c1}\right) \quad \text{and} \quad \left(x', y'\right) = \left(x'_{c1}, y'_{c1}\right)$$

(1.59)

$$F_{21} = \left(n_{fx}\right)_2 - \left[R_{11}\left(n'_{tx'}\right)_2 + R_{12}\left(n'_{ty'}\right)_2 - R_{13}\left(n'_{tz'}\right)_2\right]$$

$$@ \left(x, y\right) = \left(x_{c2}, y_{c2}\right) \quad \text{and} \quad \left(x', y'\right) = \left(x'_{c2}, y'_{c2}\right)$$

(1.60)

$$F_{22} = \left(n_{fy}\right)_2 - \left[R_{21}\left(n'_{tx'}\right)_2 + R_{22}\left(n'_{ty'}\right)_2 - R_{23}\left(n'_{tz'}\right)_2\right]$$

$$@ \left(x, y\right) = \left(x_{c2}, y_{c2}\right) \quad \text{and} \quad \left(x', y'\right) = \left(x'_{c2}, y'_{c2}\right)$$

(1.61)

Eq. (1.35a) indicates that the stiffness matrix is the partial differential of each component of the load vector with respect to each of the independent variables of the system. This leads to the development of (22 × 22) and (17 × 17) stiffness matrices for the two-point contact and one-point contact situations, respectively. Expressions for the elements forming these matrices are lengthy; they are listed in Reference 7.

1.4 Model Calculations

In a test situation, a sudden impact load was applied at the tibial center of mass with the knee straight. This dynamic load was applied in a posterior direction perpendicular to the tibial mechanical axis. Impact was simulated using sinusoidally decaying forcing pulses with different durations and different magnitudes. Each pulse is expressed as:

$$F_e(t) = A \, e^{-4.73\left(\frac{t}{t_o}\right)^2} \sin\left(\frac{\pi t}{t_o}\right)$$

(1.62)

where A and t_o are amplitude and pulse duration, respectively. Forcing pulses of this form can be simulated experimentally[93] and have been used previously as typical representations of the dynamic load in head impact analysis.[46] Figs. 1.4 and 1.5 show the sinusoidally decaying forcing pulses of constant duration and constant amplitude, respectively, that have been used in the present simulation. Sample results showing the effects of varying pulse magnitude and pulse duration on knee flexion, varus-valgus rotations, internal-external tibial rotations, linear and angular velocities of the tibia, ligamentous forces, and magnitude and location of tibio-femoral contact forces are presented here.

In the analysis, the tibia was assumed to begin its motion from rest ($v_x = v_y = v_z = \omega_\alpha = \omega_\beta = \omega_\gamma = 0$ at $t = t_0 = 0$ s) while the knee was fully extended ($\alpha = 0°$, $\beta = 90°$, $\gamma = 0°$ at $t = t_0 = 0$ s). It was found that the behavior of the system is very sensitive to the location of the initial contact points which required using double precision while performing the computations. On the other hand, even a large error in the initial values of the magnitude of the lateral and medial contact forces did not have an effect on the system's stability. This is because the behavior of a DAE system is very sensitive to unbalance only in its

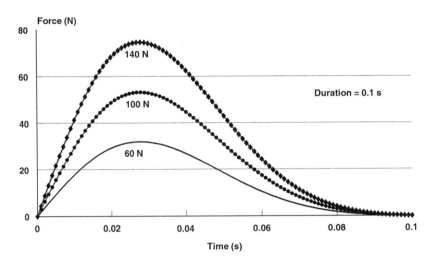

FIGURE 1.4   Forcing pulses with different amplitudes and a constant duration of 0.1 s.

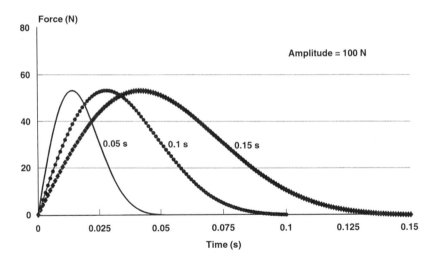

FIGURE 1.5   Forcing pulses with different durations and a constant amplitude of 100 N.

algebraic part and not in its differential part. Since neither medial nor lateral contact forces appear in the algebraic part of the DAE, the system is not sensitive to errors in their initial values.

The initial values of the x and y coordinates of the medial contact point in the local femoral and tibial coordinate systems of axes were taken as $x_c = -16.72913866466$ mm and $y_c = -18.540280082863$ mm; and $x_c' = -16.980955$ mm and $y_c' = -12.8$ mm, respectively. The initial values of the x and y coordinates of the lateral contact point in the local femoral coordinate system were taken as $x_c = 16.50009371$ mm and $y_c = -16.00200022911$ mm. The medial and lateral contact forces at $t = t_0 = 0$ s were assumed as 150.0 N and 130.0 N, respectively.

Using these assumed values, the coordinates of the tibial origin with respect to the femur at $t = t_0 = 0$ s were calculated using Eq. (1.88), thus satisfying the contact condition at the medial side. The initial values of the x and y coordinates of the lateral contact point in the tibial coordinate system were then calculated using Eq. (1.8) one more time, hence satisfying the contact condition at the lateral side. The unbalance of the system (residual) at $t = t_0 = 0$ s is then determined using Eqs. (1.40) through (1.61). An iterative procedure is then employed to determine the initial values of the 22 system variables that minimize the initial residual.

Fig. 1.6 shows how the knee flexion angle changes with time for a constant pulse amplitude of 100 N. Increasing pulse duration caused the motion to become faster. This also occurred with increasing pulse amplitude. Figs. 1.7 and 1.8 show how the varus-valgus rotation and the tibial axial rotation angles change with knee flexion, respectively, for pulses with different duration and a constant amplitude of 100 N. The one-point contact model was involved for joint positions with flexion angles larger than the flexion angle at point (A), marked by a star, in each curve in Fig. 1.7 and 1.8. For the rest of the figures presented here, point A is not marked for conciseness.

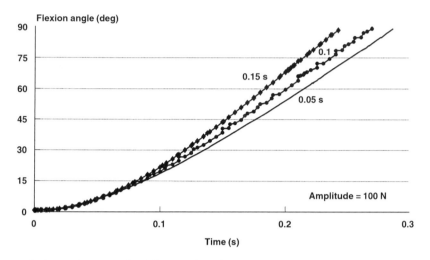

FIGURE 1.6   Flexion angle vs. time for different pulses of constant amplitude of 100 N.

Fig. 1.7 shows that valgus rotation increased in the first 10° of knee flexion, decreased to zero between 25 and 45° of knee flexion, then varus rotation increased until the knee reached about 60° of flexion, then decreased until 90° of knee flexion. The results indicate that the position at which the knee had zero degrees of varus-valgus rotation changed from 25° of knee flexion to 45° of knee flexion when the pulse duration was increased. Also, Fig. 1.7 shows that increasing the pulse duration caused a decrease in the maximum amount of varus rotation. It was further found (not shown here) that increasing pulse amplitude had the same effects as increasing pulse duration.

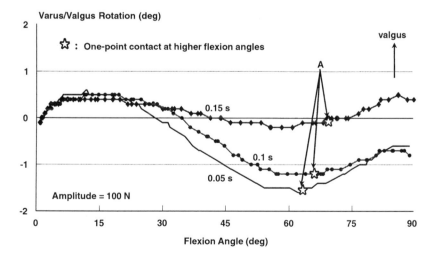

FIGURE 1.7   Varus-valgus rotations vs. flexion angle.

Fig. 1.8 shows that in the early stage of flexion, the tibia was externally rotated. This external rotation increased until it reached a maximum when the knee was between 15 and 20° of knee flexion. At this position, the knee started to go into internal rotation and reached a maximum at 90° of knee flexion. The computer simulation indicates that increasing the pulse duration (and/or amplitude) produced an increase in the magnitude of the maximum external and maximum internal rotation angles that occur during knee flexion. Also, positions of the maximum external rotation angles were slightly affected by the pulse amplitude and/or duration.

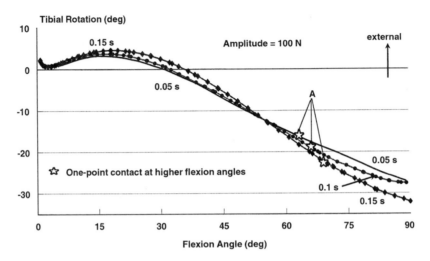

FIGURE 1.8   Tibial rotations vs. flexion angle.

Figs. 1.9 and 1.10 show the the femoral and tibial contact pathways, respectively, when a forcing pulse of 100 N amplitude and 0.1 s duration was applied to the tibia. Fig. 1.9 shows that the medial and lateral femoral contact points moved posteriorly and proximally as the knee was flexed from 0 to 90° of knee flexion. A two-point contact condition was maintained until about 66° of knee flexion. From there on, and until 90° of flexion, a one-point contact was predicted on the medial side. Fig. 1.10 shows that the medial tibial contact point moved anteriorly, while the lateral tibial contact point moved posteriorly as the knee was flexed from full extension. This motion is expected and can be thought of as a result of the femur rotating externally over fixed plateaus which causes the medial tibial contact point to move anteriorly and the lateral tibial contact point to move posteriorly. The analysis show that the position of separation in the lateral compartment was slightly affected by the amplitude and/or duration of the forcing pulses. However, the motion pattern of the medial and lateral femoral and tibial contact points was independent of both pulse amplitude and pulse duration.

Figs. 1.11 through 1.15 show how knee translational and rotational (angular) velocities changed with knee flexion for pulses with different durations and a constant amplitude of 100 N. Fig. 1.11 shows that the lateral shift velocity increased from zero at full extension to a maximum at about 15° of knee flexion, then decreased, reaching zero around 30° of knee flexion. Henceforth, the medial shift velocity increased, reaching a maximum at 90° of knee flexion. Fig. 1.12 shows that the posterior velocity increased from zero at full extension to a maximum at about 50 to 60° of knee flexion, then decreased slightly as the knee flexion increased. Fig. 1.13 shows that the knee flexion angular velocity increased monotonously from zero at full extension, reaching a maximum at 90° of knee flexion. Fig. 1.14 shows that the valgus velocity increased from zero at full extension to a maximum at about 5° of knee flexion, then decreased, reaching zero around 10° of knee flexion. Then, the varus velocity increased, reaching a maximum between 40 and 50° of knee flexion; henceforth, the velocity decreased reaching zero between 60 and 65° of knee flexion. The valgus velocity increased again achieving a maximum around 80° of knee flexion. Fig. 1.15 shows that the internal rotation velocity increased from zero to a maximum at 2° of knee flexion

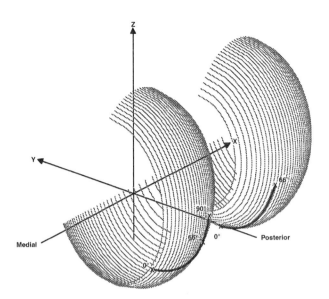

FIGURE 1.9 Femoral contact pathways. (*Source*: Rahman, E.M. and Hefzy, M.S., *JJBE: Med. Eng. Physics*, 20, 4, 276, 1998. With permission from Elsevier Science.)

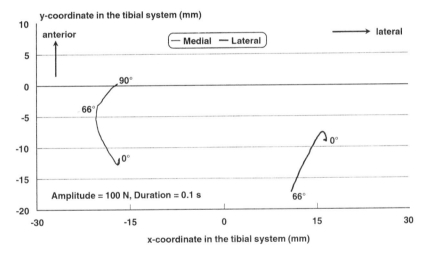

FIGURE 1.10 Tibial contact pathways. (*Source*: Rahman, E.M. and Hefzy, M.S., *JJBE: Med. Eng. Physics*, 20, 4, 276, 1998. With permission from Elsevier Science.)

and decreased to zero at 5° of knee flexion. Then, the external rotation velocity began to increase reaching a maximum at around 8° of knee flexion and decreased, reaching zero around 20° of knee flexion. From this point, the internal rotation velocity increased to a maximum between 45 and 60° of knee flexion then decreased as the knee flexion increased.

The remaining results related to contact and ligamentous forces are shown for pulses of different amplitudes and a constant duration of 0.1 s. Fig. 1.16 shows that the medial contact force decreased in the first 15° of knee flexion, then increased until 60° of knee flexion, and then decreased until 90° of knee flexion. Fig. 1.17 shows that the lateral contact force decreased from a maximum at full extension until it reached zero when separation occurred in the lateral compartment. These two figures show that increasing the pulse amplitude caused a decrease in the magnitude of the medial and lateral contact forces; similar results were obtained when the pulse duration was increased while the pulse amplitude was kept unchanged.

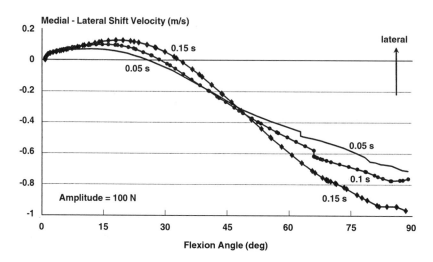

FIGURE 1.11 Medial-lateral shift velocity vs. flexion angle.

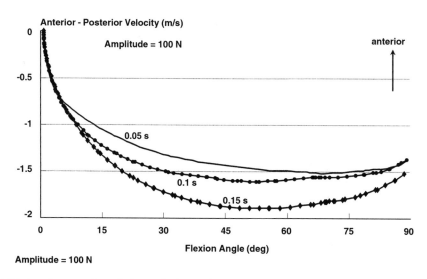

FIGURE 1.12 Anterior-posterior velocity vs. flexion angle.

Figs. 1.18 through 1.25 show how the ligamentous forces changed with knee flexion. Figs. 1.18 and 1.19 show that in the first 20° of knee flexion, the tension in the anterior cruciate ligament is greatest in its posterior fibers. As the flexion angle increased, this tension decreased while tension in the anterior fibers increased and became dominant. Figs. 1.20 and 1.21 show that the anterior fibers of the posterior cruciate ligament carried large forces after 45° of knee flexion while the posterior fibers were in tension only in the first 10° of knee flexion.

Figs. 1.22, 1.23, and 1.24 show that within the medial collateral ligament, MCL, the deep fibers carried more forces than the anterior fibers, which carried more forces than the oblique fibers which, in turn, carried relatively very small forces. The maximum forces in the anterior and deep fibers occurred between 40 and 50° of knee flexion, while the maximum force in the oblique fibers occurred at approximately 5° of knee flexion. Fig. 1.25 shows that the lateral collateral ligament sustained forces only in the first 45 to 60° of knee flexion, with a maximum value occurring near full extension.

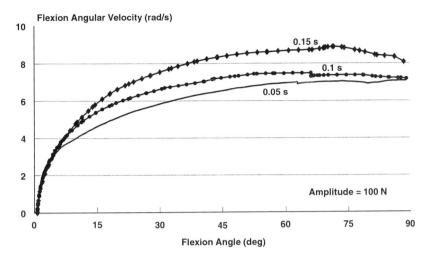

FIGURE 1.13 Knee flexion angular velocity vs. flexion angle.

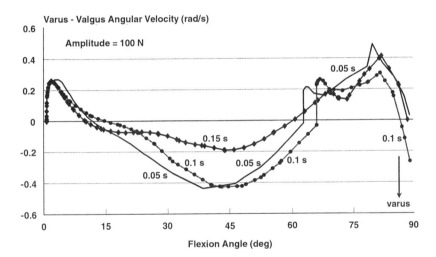

FIGURE 1.14 Varus-valgus angular velocity vs. flexion angle.

The results show that the patterns of change in the ligamentous forces were not generally affected by changing the characteristics of the applied pulsing loads. However, increasing pulse amplitude (and/or duration) slightly affected the magnitude of the forces in the different ligamentous fibers.

1.5 Discussion

Review of the literature reveals that most of the published anatomical dynamic knee models are two-dimensional[1-3,47-49,84,93-96] and that most of the existing three-dimensional anatomical knee models are quasi-static in nature.[9,20-23,50,129,130] Quasi-static models determine forces and motion parameters by solving the equilibrium equations, subject to appropriate constraints, at a specific knee position. The procedure is then repeated at different positions to cover a range of knee motions. However, these quasi-static models cannot predict the velocity or acceleration of the different segments forming the joint. Also, these models are further limited in that they cannot determine the effects of the dynamic inertial loads (which occur in many daily living activities) on joint kinematics and joint loads. In this chapter, a

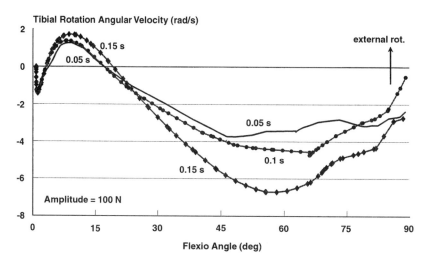

FIGURE 1.15 Tibial rotation angular velocity vs. flexion angle.

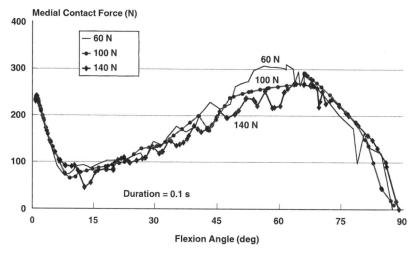

FIGURE 1.16 Medial tibio-femoral contact force vs. flexion angle. (*Source*: Rahman, E.M. and Hefzy, M.S., *JJBE: Med. Eng. Physics*, 20, 4, 276, 1998. With permission from Elsevier Science.)

three-dimensional dynamic anatomical model of the tibio-femoral joint that predicts its response under sudden impact loads is presented.

The system of equations forming an anatomical quasi-static knee model is a system of nonlinear algebraic equations. These equations are solved iteratively using a Newton-Raphson iteration technique,[20-23,129,130] discretized and solved using the finite element method[9] or rewritten as a potential energy function that can be minimized using an optimization method such as the steepest descent optimization technique.[50] On the other hand, the system of equations forming an anatomical dynamic model is a system of differential algebraic equations (DAEs). Solving a DAE system is more difficult than solving an algebraic system. Several techniques have been proposed to solve the DAE system that describes the two-dimensional dynamic response of the knee joint.[1-3,47,84,93-96,118-119] These techniques were limited in that they could not solve the DAE system that represents the three-dimensional situation. Using the Differential/Algebraic System Solver software (DASSL) developed at the Lawrence Livermore National Laboratory, the latter and more complex DAE system was solved, thus describing the three-dimensional dynamic response of the knee joint. The integration scheme implemented in DASSL employs variable order and variable size multistep backward differentiation formulas (BDFs).

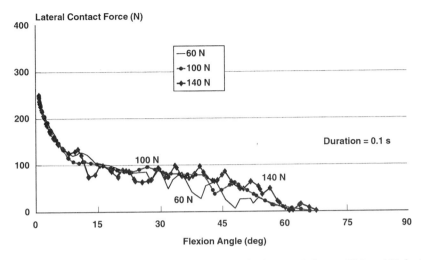

FIGURE 1.17 Lateral tibio-femoral contact force vs. flexion angle. (*Source*: Rahman, E.M. and Hefzy, M.S., *JJBE: Med. Eng. Physics*, 20, 4, 276, 1998. With permission from Elsevier Science.)

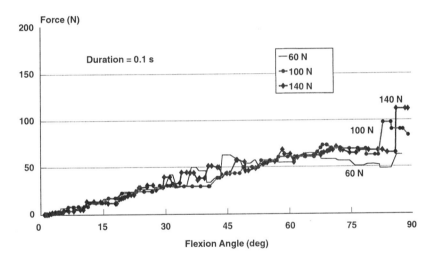

FIGURE 1.18 Forces in the anterior fibers of the anterior cruciate ligament.

It is hard to validate the present model predictions because of the limited experimental data available in the literature that describe the dynamic behavior of the human knee joint. The dynamic response of the joint must be described in terms of the loads exerted on the joint; the six components of the three-dimensional motion of the tibia with respect to the femur; the deformations of the different components forming the joint, including the ligaments, menisci and cartilage. Dortmans et al.[42] described the difficulties associated with this task by stating that the dynamic behavior "... is much too complicated to deal with, due to a lack of experimental techniques to quantify all parameters." Keeping this in mind, model predictions are discussed and, whenever appropriate, compared with the experimental data in the literature.

Varus-Valgus Rotation

Model predictions show that varus rotation occurred in association with internal tibial rotation while the knee was flexed. These results are in agreement with van Kampen et al.[79] who reported that applying internal torque to the tibia produced varus rotation. They are also in agreement with Essinger et al.'s model calculations where varus rotations and internal tibial rotations were predicted with knee flexion.[50]

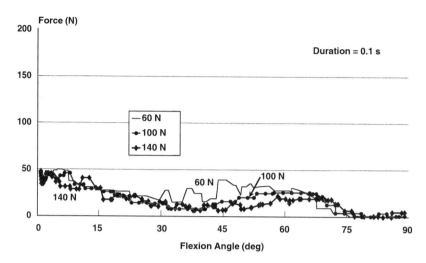

FIGURE 1.19 Forces in the posterior fibers of the anterior cruciate ligament.

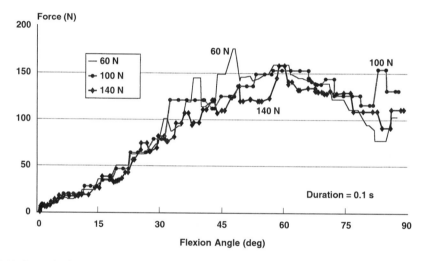

FIGURE 1.20 Forces in the anterior fibers of the posterior cruciate ligament.

Also, the results reported here are within the varus-valgus rotation's envelope of passive knee motion reported by Blankevoort et al.[19,21,23] and Mills and Hull.[91,92] The envelope of passive knee motion is the region where the joint offers little resistance to motion. However, the model predictions are not in agreement with the data reported by Blankevoort et al.[19,21,23] and Mills and Hull[91,92] where a valgus rotation was coupled with and caused by the internal tibial rotation. This difference may be due to the omission of the menisci in this model.

Mills and Hull[92] reported that the valgus rotation coupled with the internal tibial rotation is due to the medial condyle's ride over the medial meniscus. When an internal moment of 3 N-m was applied by van Kampen et al.[79] to the tibia (significantly lower than the 20 N-m applied by Mills and Hull[91,92] which did not force the ride of the femoral condyle over the medial meniscus), a varus rotation occurred with knee flexion. Both the present model and Essinger et al.'s model[50] neglected the menisci and predicted varus rotations associated with internal tibial rotation and knee flexion.

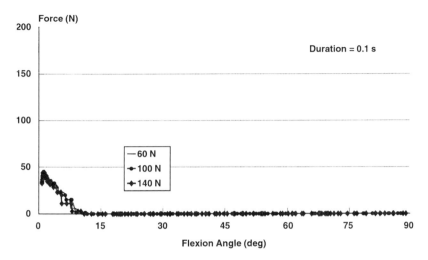

FIGURE 1.21 Forces in the posterior fibers of the posterior cruciate ligament.

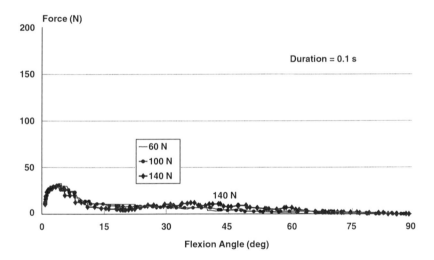

FIGURE 1.22 Forces in the oblique fibers of the medial collateral ligament.

Tibial Rotation

Model calculations show that as the knee was flexed from 15 to 90°, it underwent internal tibial rotation. This rotation indicates that the tibia was subjected to internal moments caused by tension forces in the lateral collateral ligament and/or the anterior fibers of both the anterior and posterior cruciate ligaments. The predicted tibial rotations lay within the envelope of passive knee motion defined earlier.[19,21,23] Further, these results are in agreement with those of Nigg et al.[103] who reported a mean internal tibial rotation of 21.8° ± 8.4° during walking (that is, a range of 0 to 70° of knee flexion) and those of Essinger et al.[50] who reported an internal tibial rotation of 15° when the knee was flexed from full extension to 90° of knee flexion.

These results indicate that external tibial rotation occurs with knee extension. However, this pattern was not predicted by the model in the first 15° of knee flexion. This finding is important since it is in agreement with the emerging thought of the need to re-evaluate the so-called screw-home mechanism during dynamic motions. The "screw-home mechanism" is described as the external rotation of the tibia with respect to the femur that occurs during the terminal stages of knee extension.[100] This mechanism

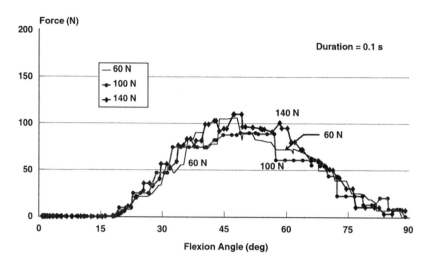

FIGURE 1.23 Forces in the anterior fibers of the medial collateral ligament.

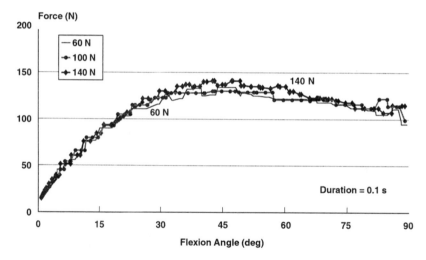

FIGURE 1.24 Forces in the deep fibers of the medial collateral ligament.

was not predicted by this model since it was found that the tibia rotated internally as the knee was extended from 15° to full extension. These results are in agreement with the recent data reported by Blankevoort et al.[19] and Lafortune et al.[82] Lafortune et al. found that the tibia rotated internally with respect to the femur in the terminal stage of knee extension. They stated that their results "appear to call into question the accepted view that the tibia rotates externally relative to the femur in the later stages of knee extension." Blankevoort et al.[19] also reported that this screw-home mechanism was not present when passive joint motion characteristics were determined.

Femoral and Tibial Contact Pathways

Model calculations show that the femoral contact points moved posteriorly and proximally with knee flexion on both the medial and lateral femoral condyles over the whole range of knee flexion. There was almost no medial-lateral shift on the femoral condyles. These results are in agreement with those obtained using the two-dimensional four-bar linkage model[100,101] and those reported using two-dimensional anatomical dynamic models.[1-3] These results are also in agreement with the data in the literature describing

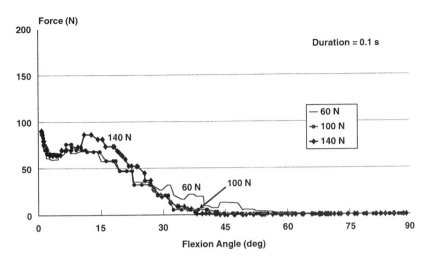

FIGURE 1.25 Forces in the lateral collateral ligament.

the three-dimensional tibio-femoral contact characteristics where the tibio-femoral contact areas moved posteriorly and proximally over the femoral condyles with knee flexion.[23,123]

The medial tibial contact point was found to move anteriorly and the lateral tibial contact point to move posteriorly when the knee was flexed beyond 15°. This is consistent with the predicted internal tibial rotation pattern that occurred in this range of knee flexion. When separation occurred in the lateral compartment, the medial contact point continued to move anteriorly to 90° of knee flexion as the internal tibial rotation continued to be observed. Within the first 15° of knee flexion, the pattern describing the pathways of the tibial contact points was reversed. The medial tibial contact point moved posteriorly, while the lateral tibial contact point moved anteriorly. This pattern of motion was due to the external tibial rotation sustained in this range of knee flexion.

Velocity of the Tibia

The posterior velocity of the tibia increased as it was flexed due to the application of the posterior load which consisted of the forcing pulse and the leg's weight. As the magnitude of the forcing pulse approached zero and the component of the weight along the posterior axis decreased with knee flexion, the posterior velocity leveled off to almost a constant magnitude for the remainder of the motion. The flexion angular velocity also increased with flexion and time, reaching a maximum at 90° of knee flexion. Both posterior velocity and flexion angular velocity of the tibia increased with increasing forcing pulse amplitude and/or duration.

Combining model predictions for knee kinematics provides a better understanding of the three-dimensional knee motions for the conditions tested. As the knee rotated in valgus, the valgus velocity increased from rest reaching a maximum around 5° of knee flexion, then decreased to zero around 10° of knee flexion. As the motion changed direction and the knee began to rotate in varus, the varus velocity increased reaching a maximum around 45° of knee flexion. While the varus velocity decreased after that, the tibia continued to rotate in varus until separation in the lateral compartment occurred, then varus rotation stopped and the varus velocity reached zero. At this point, the knee began to rotate back toward a zero varus-valgus position and the valgus velocity increased henceforth until 90° of knee flexion.

Furthermore, due to the varus-valgus rotation, the leg acts as a pendulum in the frontal plane. As the valgus rotation of the knee increased during flexion from full extension (Fig. 1.7), the tibial center of gravity developed a lateral shift velocity (Fig. 1.11). The lateral shift velocity reached a maximum around 15° of knee flexion as the valgus angulation reached a maximum. At this point, the leg began to rotate in varus and the lateral shift velocity started to decrease until it reached zero around 30° of knee flexion. A medial shift velocity was predicted from this point on and continued to increase throughout knee motion.

Magnitude of the Tibio-Femoral Contact Forces

Model predictions show that increasing the pulse amplitude and/or duration caused a decrease in the magnitude of the contact forces. This was associated with an increase in the flexion velocity. This is consistent with the results reported by Kaufman et al.[81] who calculated muscle and joint forces in the knee during isokinetic exercise. They found that joint loads were higher at lower flexion velocities than they were at higher flexion velocities.

It was also found that the maximum value of the medial contact force was larger than the maximum value of the lateral contact force. Maximum medial contact force occurred when the knee was in maximum varus angulation. These results are in agreement with those in the literature reporting that the medial condyles carry more load than the lateral.[33,98,123] The maximum medial and lateral contact forces reported here are lower than those reported in the literature.[33] This is expected since this model does not include ground reaction forces nor muscle forces, both of which produce high joint contact forces.[112]

As indicated earlier, limited experimental data are available in the literature describing the dynamic behavior of the human knee joint including contact characteristics. Most of this work was conducted by Jans et al.[78] and Dortmans et al.[42,43] The results obtained from this model describing joint contact are different in nature from their results which were presented in terms of a transfer function between the applied loads and the displacements of the tibia. Yet, the model predictions are indirectly in agreement with their results. They reported a decrease of the joint stiffness with increasing load level. The decrease in the contact force reported here, which is associated with an increase in the amplitude and/or duration of the forcing pulse, indicates a reduction in the joint stiffness that occurs with an increase in the load level.

Ligamentous Forces

Almost all of the data available in the literature (experimental or analytical using mathematical models) that describe ligament function are static or quasi-static in nature. Hence, it is hard to compare the ligamentous forces predicted using the present model with those reported elsewhere because the dynamic response is much different from the static response of the joint.[135] Moreover, review of the literature reveals many differences of opinion with regard to knee ligament function.[45,52] There are several reasons for these discrepancies in results and conflicts of opinions, depending on whether ligamentous forces are determined experimentally or predicted analytically using a mathematical model.

Experimentally, placing a strain measurement device at different locations on the same ligament will display different strain patterns because different fiber bundles within a ligament function differently through the range of knee motion.[10,52,65,128] Reported findings related to ligament function are also affected by various other factors including

1. The use of qualitative or indirect methods, such as palpation.[45]
2. The use of methods that interfere with the loading patterns of the ligaments, such as bulky strain gauges[45,52] or methods that prestrain the ligaments, such as buckle transducers.[8,11,107]
3. The use of very compliant strain transducers, such as the Hall effect strain transducer and the liquid metal strain gauge transducer, leading to erroneous data, such as compressive axial tissue strain.[18,59]
4. The use of experimental protocols that change the relationships between the loading patterns of the ligaments, such as cutting one of the ligaments and reporting the results of the other ligamentous structures.[10,45,52]

Analytically, model predictions depend on the values of the stiffness coefficients of the spring elements representing the ligaments. Different factors must be considered when these coefficients are specified using the experimental data available in the literature. For instance, it has been reported that the material properties of the ligamentous structures are location-dependent.[26,29] The actual stiffness of soft tissue in midsubstance is two or three times higher than that recorded from grip to grip due to slippage and stress concentration at the grip.[26] It has also been reported that ligament strength depends on loading orientation. Available experimental data have shown that a direct relationship exists between the orientation of the ligament with respect to the applied force and its strength.[27,51,111,117,132,133] As a result, the strength of a ligament varies with the flexion angle at which the force-displacement test is performed.

Figgie et al.[51] and Butler et al.[27] related this behavior to the nonhomogeneous fascicle organization within the ligamentous structure which causes the strength of the ligament to increase as more fascicles become aligned with the loading force. In this mathematical model, ligaments were divided into ligament bundles to account for macro-differences in orientation within ligaments. However, micro-differences within individual bundles were not considered since data in the literature are insufficient to quantify the stiffnesses of the different fiber bundles at different flexion angles. In the following, and within these qualifications, predicted ligamentous forces will be discussed and compared with those available in the literature.

Model predictions indicate that increasing the amplitude and/or duration of the forcing pulse does not change the load sharing relations between the different ligamentous structures. This result was expected since the forces in a ligament depend essentially on its length, which is a function of the relative position of the tibia with respect to the femur. Thus, as long as it does not change the pattern of the tibia's translation with respect to the femur, an increase in the force applied to the tibia will not change the load sharing relations between the ligaments.

The anterior and posterior fibers of the anterior cruciate ligament (ACL) had opposite force patterns. The anterior fibers of the ACL were slack at full extension and tightened progressively as the knee was flexed reaching a maximum of 70 N at 90° of knee flexion. The posterior fibers of the ACL were most taut at full extension, carrying a load of 50 N; the tension decreased until it vanished around 75° of knee flexion. These data show that the anterior portion of the ligament is shorter at full extension and longer at 90° of knee flexion, while the posterior portion is longer at full extension and shorter at 90° of flexion. These results are in agreement with those obtained from the different qualitative,[58] length patterns,[20,65,117,126] and direct force measurement studies[52] reported in the literature to describe the functions of the anterior and posterior fibers of the ACL. The predicted maximum forces of 70 N and 50 N in the anterior and posterior fibers, respectively, are also in agreement with those reported in the literature. Using strain gauges, France et al.[52] measured the forces in these fibers and reported a maximum force of 68 N in the anterior fibers at 70° of knee flexion and a maximum force of 40 N in the posterior fibers at full extension.

The forces in the anterior fibers of the posterior cruciate ligament (PCL) increased from zero at full extension to a maximum of 150 N around 60° of knee flexion, and then decreased until 90° of knee flexion. On the contrary, the forces in the posterior fibers of the PCL were maximum, carrying a load of 50 N at full extension and reached zero around 10° of knee flexion. These data show that the anterior portion of the ligament is shorter at full extension and longer at 90° of knee flexion, while the posterior portion is longer at full extension and shorter at 90° of flexion. The predicted maximum forces of 150 and 50 N in the anterior and posterior fibers, respectively, are higher than those reported in the literature.[52] Yet, these results are in agreement with the data available in the literature indicating that the PCL bundles have reciprocal tightening and slackening patterns in flexion and extension. In flexion the anterior fibers are tight and the posterior fibers are slack; in extension this trend is reversed.[20,41,52,58,62,74,104,109]

The medial collateral ligament (MCL) was discretized into anterior, deep, and oblique fibers, thus neglecting the posterior fibers. This assumption was introduced, considering that the line segment representation of fibers (used in the present analysis) is not adequate to model the posterior fibers since they wrap around the medial condyle. It was found that the forces in the oblique fibers of the MCL were maximum near full extension where they carried a load of 30 N and decreased with knee flexion, with very little force in the fibers beyond 20° of knee flexion. Forces in the anterior fibers of the ligament were almost zero in the first 20° of knee flexion. With more flexion, these forces increased to a maximum of 100 N at around 50° of knee flexion, then decreased, reaching zero at 90° of knee flexion. Forces in the deep fibers increased with knee flexion from 0°, reaching a maximum of 90 N around 45° of knee flexion, then remained almost constant to 90° of knee flexion. It was found that the anterior and deep fibers of the MCL carried most of the load within the ligament. Forces in the oblique fibers were relatively small. These results are in agreement with the data reported in the literature which indicate that the anterior fibers are longest at around 50° of flexion[10,36,128] and the oblique fibers are longest in extension.[20,36]

In the model developed by Essinger et al.,[50] the MCL was divided into anterior and posterior elements. The force in the anterior element attained a maximum value of 250 N between 60 and 70° of knee flexion, which is a much higher value than the value predicted by the present model of 100 N. This is probably because the deep fibers were not considered as a separate entity in Essinger et al.'s model. This caused the force in the MCL to be distributed among fewer elements, thus producing higher forces in each of these elements.

The force in the lateral collateral ligament (LCL) was at a maximum of 90 N, at full extension, and decreased with knee flexion until it reached a very small value around 35° of knee flexion. These results are in agreement with the results available in the literature indicating that the LCL attains its greatest length at extension and becomes progressively shorter with flexion.[36,45,50,52,104,126]

1.6   Conclusions

This chapter presents a three-dimensional anatomical dynamic model of the tibio-femoral joint that predicts its response under sudden impact loads. Model calculations suggest that the three-dimensional dynamic anatomical modeling of the human musculo-skeletal joints is a versatile tool for the study of the internal forces in these joints. Results produced by such anatomical models are more useful in studying the responses of the different structures forming these joints than those obtained using less sophisticated models because these anatomical models can account for the dynamic effects of the external loads, the anatomy of the joints, and the constitutive relations of the force-contributing structures. In the formulation presented here, all the coordinates of the ligamentous attachment sites were dependent variables. As a result, it is possible to introduce more ligaments and/or split each ligament into several fiber bundles. This formulation allowed solution of the three-dimensional model which could not be solved using Moeinzadeh's formulation[96] a decade ago.

The results obtained from this study describing the three-dimensional knee motions indicate a need to re-evaluate the "screw-home mechanism" which calls for external tibial rotation in the final stages of knee extension. This mechanism was not predicted by the model since it was found that the tibia rotated internally as the knee was extended from 20° of knee flexion to full extension.

Furthermore, evidence of symmetric "femoral roll back" was not observed from the model predictions. In the range of 20 to 66° of knee flexion, the medial tibial contact point moved anteriorly while the lateral tibial contact point moved posteriorly. In the range of 66 to 90° of knee flexion, contact was maintained only on the medial side and the tibial contact point (on the medial side) continued to move anteriorly.

It was found that increasing the pulse magnitude caused a decrease in the magnitude of the contact forces and an increase in the magnitude of both tibial linear and angular velocities. This decrease in the magnitude of the contact forces reflects a decrease in joint stiffness. These results are consistent with Kaufman's data[81] reporting that a decrease in knee stiffness is associated with an increase in the speed of the tibia.

Model predictions also show that the major fiber bundles resisting a posterior forcing pulse acting on the tibia in the range of 20 to 90° of knee flexion are the anterior fibers of the PCL and the anterior and deep fibers of the MCL explaining the clinical finding reported by Loos et al.[85] and Cross and Powell[35] that most of isolated PCL injuries or combined injuries to the PCL and the MCL result from car dashboard injuries, motorcycle accidents, or falls on flexed knees all of which produce posterior tibial impacts on flexed knees, causing these isolated or combined injuries.

Furthermore, the results show reciprocal load patterns in the anterior and posterior fibers of the ACL and PCL. The results indicate that both anterior and posterior fibers of the ACL carry significant loads in a large range of knee motion. They also show that the posterior fibers of the PCL carry small loads over a small range of motion, from full extension to 10° of knee flexion. These results suggest that under the conditions tested, it could be enough to reconstruct the anterior fibers of the PCL to regain the stability of a PCL deficient knee. On the other hand, regaining stability of an ACL deficient knee would require the reconstruction of both the anterior and posterior fibers of the ACL since both fiber bundles carry significant loads for a large range of knee motion.

1.7 Future Work

The initial results presented in this work encourage us to consider this model as a versatile tool for the study of the internal forces within the knee joint. Future work includes incorporating the patella into the model and using it to determine knee response under different loading conditions and to predict the behavior of the joint following ligamentous injuries and different reconstruction procedures. The introduction of the patella into the present model can be achieved by modeling the knee using three rigid body segments that comprise the tibio-femoral and patello-femoral joints. Some of the forces acting on the tibia and patella are shown in Fig. 1.26. These two bones are connected by an extensible link that represents the patellar tendon. Similar to the present tibio-femoral model, the femur will be assumed fixed, and the friction forces between the femur and patella will be neglected.

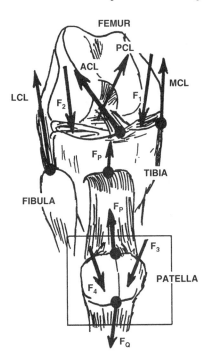

FIGURE 1.26 A general three-dimensional knee model that includes tibio-femoral and patello-femoral joints. The force in the patellar tendon, F_P, acts as a coupling force between the system of equations describing tibio-femoral motions and the system of equations describing patello-femoral motions.

In the kinematic analysis, a third coordinate system will be identified on the moving patella. The patello-femoral joint coordinate system[67] will be used to describe patello-femoral motions in terms of six kinematic parameters: three rotations (patellar flexion-extension, patellar medial-lateral tilt, and patellar rotation) and three translations (medial-lateral patellar shift, anterior-posterior patellar translation, and proximal-distal patellar translation).

When writing the equations of motion, two subsystems will be identified: tibio-femoral and patello-femoral systems. The force in the patellar tendon, F_P, will be assumed to act as a coupling force between these two systems. The tendon will be considered a rigid ligament whose length remains constant during motion. This rigid patellar ligament condition is expressed mathematically as one equation. The force in this ligament is the coupling force between the two subsystems of equations describing the tibio-femoral and patello-femoral joint motions. This force, $\vec{F}_p$, is expressed as:

$$\vec{F}_p = F_p \left(\vec{R}_a - \vec{R}_t \right) \Big/ \left\| \left(\vec{R}_a - \vec{R}_t \right) \right\| \tag{1.63}$$

where F_p is the magnitude of the patellar ligament force and $\vec{R}_a$ is the position vector of the tibial tuberosity in the femoral coordinate system expressed in terms of its local tibial coordinates and the six unknown kinematic parameters describing the tibio-femoral motions. Similarly, $\vec{R}_t$ is the position vector of the patellar apex in the femoral coordinate system expressed in terms of its local patellar coordinates and the six unknown kinematic parameters describing the patello-femoral motions.

The patello-femoral contact will be modeled by assuming that a two-point frictionless contact exists at all times on the medial and lateral sides such that four forces act on the patella at any instant: the force exerted by the quadriceps muscle, the force in the patellar ligament, and the medial and lateral contact forces acting on the medial and lateral patellar facets. The patella will be assumed massless; accordingly, patellar equations of motion reduce to six equilibrium equations.

An analysis similar to that of the tibio-femoral contact will be employed. The position vectors of each of the two contact points in the femoral and patellar coordinate systems will be related using the rotation matrix defined in terms of the six unknown kinematic parameters that describe patello-femoral motions. Writing this relation at each of the two contact points generates six scalar equations which represent the patello-femoral contact conditions. Furthermore, Eq. (1.3) is also used to enforce the rigid bodies condition which is expressed mathematically as four equations that represent the geometric compatibility conditions for the patello-femoral joint.

The system of equations describing patello-femoral motions will thus consist of 17 equations: six equilibrium equations, six patello-femoral contact conditions, four patello-femoral compatibility conditions and one rigid patellar ligament condition. Combining both systems we obtain, for the two-point tibio-femoral contact situation, a system of 33 differential algebraic equations in the following 33 unknowns: (1) six motion parameters describing tibio-femoral joint motions; (2) six motion parameters describing patello-femoral joint motions; (3) eight parameters representing the x and y coordinates of each of the two tibio-femoral contact points in both femoral and tibial systems; (4) eight parameters representing the x and y coordinates of each of the two patello-femoral contact points in both patellar and femoral systems; (5) two parameters representing the magnitude of the tibio-femoral contact forces; (6) two parameters representing the ratios of the two patello-femoral contact forces to the quadriceps tendon force; and (7) one parameter describing the ratio of the patellar tendon force to the quadriceps tendon force.

As a first step, and in order to simplify the solution algorithm of this general model (which involves solving 33 nonlinear differential algebraic equations), an iterative and approximate procedure can be adopted in which the tibio-femoral and patello-femoral systems of equations can be solved concurrently. The solution consists of finding the position of the patella for a given tibial position. Thus, the nonlinear differential algebraic equations describing the tibio-femoral system will be solved first. The nonlinear algebraic equations describing the equilibrium of the patella will then be solved to determine the position of the patella along with the patellar ligament force which acts as the coupling force between the tibio-femoral and patello-femoral systems.

In this approximate procedure, the initial conditions require the specification of the six kinematic parameters describing the tibio-femoral motions, and the eight parameters specifying the local x and y coordinates of the medial and lateral contact points in both femoral and tibial coordinate systems of axes. These initial values must satisfy the tibio-femoral contact and compatibility equations. Knowing the initial tibio-femoral position, the initial position of the tibial tuberosity with respect to the femoral origin can be calculated and then used in conjunction with the quadriceps muscle force as part of the patello-femoral input data to solve the patello-femoral system of equations for the initial position of the patella. The quadriceps force is an input to this system and must be specified as an external load.

The solution of the patello-femoral system of equations provides the initial position of the patellar apex with respect to the femoral origin and the initial value of the patellar ligament force. The initial values of these variables are then used as input to a DAE solver to find the solution of the tibio-femoral system of equations after a time step Δt. Knowing the tibio-femoral motions after the time step Δt, the position of the tibial tuberosity with respect to the femoral origin can be calculated at the end of this time interval. This new position of the tibial tuberosity is then used as part of a new set of patello-femoral input data to solve the patello-femoral system of equations after a time step Δt at which the value of the quadriceps tendon force is evaluated from the input function at the new time station. The subsequent solution of the patello-femoral system of equations provides the position of the patellar apex with respect to the femoral origin and the value of the patellar ligament force after a time step Δt. The values of these two variables are then used as input to find the solution of the tibio-femoral system of equations after a second time step Δt, that is, at time station 2 Δt. This iterative process continues until the motion is tracked over the complete range of motion.

Acknowledgment

Part of this work was supported by grants BCS 9209078 and BES 9809243 from the Biomedical Program of the Biomedical Engineering and Environmental Systems Division of the National Science Foundation. A portion of this chapter was prepared while Dr. Hefzy was on leave at the Department of Biological and Medical Research, King Faisal Specialist Hospital and Research Centre, Riyadh, Kingdom of Saudi Arabia.

References

1. Abdel-Rahman, E., A two-dimensional dynamic model of the human knee joint, Master's thesis, University of Toledo, Toledo, OH, 1991.
2. Abdel-Rahman, E. and Hefzy, M.S., Two-dimensional dynamic model of the tibio-femoral joint, *Adv. Bioeng.*, 20, 413, 1991.
3. Abdel-Rahman, E. and Hefzy, M.S., A two-dimensional dynamic anatomical model of the human knee joint, *ASME J. Biomechanical Eng.*, 115, 357, 1993.
4. Abdel-Rahman, E. and Hefzy, M.S., Three-dimensional dynamic modeling of the tibio-femoral joint, *Adv. Bioeng.*, 26. 315, 1993.
5. Abdel-Rayman, E., Afjeh, A., and Hefzy, M.S., An improved solution algorithm for the determination of the 3-D dynamic response of the tibio-femoral joint, Proceedings of the 13th Southern Biomedical Engineering Conference, April 1994, 368.
6. Abdel-Rahman, E., Hefzy, M.S., and Afjeh, A., Determination of the three-dimensional dynamic response of the tibio-femoral joint using a DAE solver, *Adv. Bioeng.*, 28, 421, 1994.
7. Abdel-Rahman, E., A three-dimensional dynamic anatomically based model of the human tibio-femoral joint, Doctoral dissertation, University of Toledo, Toledo, OH, 1995.
8. An, K.-N., Berglund, L., Cooney, W.P., Chao, E.Y.S., and Kovacevic, N., Direct *in vivo* tendon force measurement system, *J. Biomechanics*, 23, 1269, 1990.
9. Andriacchi, T.P., Mikosz, R.P., Hampton, S.J., and Galante, J.O., Model studies of the stiffness characteristics of the human knee joint, *J. Biomechanics*, 16, 23, 1983.
10. Arms, S., Boyle, J., Johnson, R., and Pope, M., Strain measurement in the medical collateral ligament of the human knee: an autopsy study, *J. Biomechanics*, 16, 491, 1983.
11. Arms, S.W., Pope, M.H., Johnson, R.J., Fisher, R.A., Arvidsson, I., and Eriksson, E., The biomechanics of anterior cruciate ligament rehabilitation and reconstruction, *Am. J. Sports Med.*, 12, 8, 1984.

12. Ateshian, G.A., Soslowsky, L.J., and Mow, V.C., Quantitation of articular surface topography and cartilage thickness in knee joints using stereophotogrammetry, *J. Biomechanics*, 24, 776, 1991.

13. Ateshian, G.A., A B-spline least-squares surface-fitting method for articular surfaces of diarthrodial joints, *ASME J. Biomechanical Eng.*, 115, 366, 1993.

14. Ateshian, G.A., Kwak, S.D., Soslowsky, L.J., and Mow, V.C., A stereophotogrammetric method for determining *in situ* contact areas in diarthrodial joints and a comparison with other methods, *J. Biomechanics*, 27, 111, 1994.

15. Bathe, K.J., *Finite Element Procedures in Engineering Analysis*, Prentice-Hall, Englewood Cliffs, 1982.

16. Bathe, K.J. and Cimento, A.P., Some practical procedures for the solution of nonlinear finite element equations, *Computational Methods Appl. Mech. Eng.*, 22, 59, 1980.

17. Bendjaballah, M.Z., Shirazi-Adl, A., and Zukor, D.J., Biomechanics of the human knee joint in compression: reconstruction, mesh generation and finite element analysis, *Knee*, 2, 69, 1995.

18. Beynnon, B.D., Pope, M.H., Fleming, B.C., Howe, J.G., Johnson, R.J., Erickson, A.R., Wertheimer, C.M., and Nichols, C., An *in vivo* study of the ACL strain biomechanics in normal knee, *Trans. Orthopaed. Res. Soc.*, 14, 324, 1989.

19. Blankevoort, L., Huiskes, R., and deLange, A., The envelope of passive knee joint motion, *J. Biomechanics*, 21, 705, 1988.

20. Blankevoort, L., Huiskes, R., and deLange, A., Recruitment of knee joint ligaments, *ASME J. Biomechanical Eng.*, 113, 94, 1991.

21. Blankevoort, L. and Huiskes, R., Ligament-bond interaction in a three-dimensional model of the knee, *ASME J. Biomechanical Eng.*, 113, 263, 1991.

22. Blankevoort, L. and Huiskes, R., Parametric validation of a 3-D knee joint model, *J. Biomechanics*, 24, 488, 1991.

23. Blankevoort, L., Kuiper, J.H., Huiskes, R., and Grootenboer, H.J., Articular contact in a three-dimensional model of the knee, *J. Biomechanics*, 24, 1019, 1991.

24. Brenan, K.E., Campbell, S.L., and Petzold, L.R., *Numerical Solution of Initial Value Problems in Differential-Algebraic Equations*, North Holland/Elsevier, Amsterdam, 1989.

25. Biczek, F.L. and Cavanaugh, P.R., Stance phase knee and ankle kinematics and kinetics during level and down-hill running, *Med. Sci. Sports Exercise*, 22, 669, 1990.

26. Butler, D.L., Groos, E.S., Noyes, F.R., Zernicke, R.F., and Brackett, K., Effects of structure and strain measurement technique on the material properties of young human tendons and fascia, *J. Biomechanics*, 17, 579, 1984.

27. Butler, D.L., Kay, M.D., and Stouffer, D.C., Comparison of material properties in fascicle-bone units from human patellar tendon and knee ligaments, *J. Biomechanics*, 19, 425, 1986.

28. Butler, D.L., Sheh, M.Y., Stouffer, D.C., Samaranayake, V.A., and Levy, M.S., Surface strain variation in human patellar tendon and knee cruciate ligaments, *ASME J. Biomechanical Eng.*, 112, 39, 1990.

29. Butler, D.L., Guan, Y., Kay, M.D., Cummings, J.F., Feder, S.M., and Levy, M.S., Location-dependent variations in the material properties of the anterior cruciate ligament, *J. Biomechanics*, 25, 511, 1992.

30. Carlin, G.J., Morrow, D., Harner, C.D., Kusayama, T., and Woo, S.L.-Y., Mechanical properties of the human posterior cruciate ligament, Proceedings of the 13th Southern Biomedical Engineering Conference, April 1994, 887.

31. Chand, R., Haug, E., and Rim, K., Stresses in the human knee joint, *J. Biomechanics*, 9, 417, 1976.

32. Chandler, R.F., Clauser, C.E., McConville, J.T., Reynolds, H.M., and Young, J.W., Investigation of intertial properties of the human body, Report AMRL-TR-74-137, Aerospace Medical Research Laboratory, Wright-Patterson Air Force Base, OH, March 1975.

33. Cheng, C.K., A mathematical model for predicting bony contact forces and muscle forces at the knee during human gait, Ph.D. dissertation, University of Iowa, Iowa City, IA, 1988.

34. Clauser, C.E., McConville, J.T., and Young, J.W., Weight, volume, and center of mass of segments of the human body, Report AMRL-TR-69-70, Aerospace Medical Research Laboratory, Wright-Patterson Air Force Base, OH, August 1969.

35. Cross, M.J. and Powell, J.F., Long-term follow-up of posterior cruciate ligament rupture: a study of 116 cases, *Am. J. Sports Med.*, 12, 292, 1984.

36. Crowninshield, R., Pope, M.H., and Johnson, R.J., An analytical model of the knee, *J. Biomechanics*, 9, 397, 1976.

37. Crowninshield, R., Pope, M.H., Johnson, R., and Miller, R., The impedance of the human knee, *J. Biomechanics*, 9, 529, 1976.

38. Davy, D.T. and Audu, M.L., A dynamic optimization technique for predicting muscle forces in the swing phase of the gait, *J. Biomechanics*, 20, 187, 1987.

39. DeLooze, M.P., Toussaint, H.M., van Dieen, H., and Kemper, H.C.G., Joint moments and muscle activity in the lower extremities and lower back in lifting and lowering tasks, *J. Biomechanics*, 26, 1067, 1993.

40. Dempster, W.T., Space requirements of the seated operator, Report AD-087-892, Wright-Patterson Air Force Base, OH, July 1955.

41. van Dijk, R., Huiskes, R., and Selvik, G., Roentgen stereophotogrammetric methods for the evaluation of the three-dimensional kinematic behavior and cruciate ligament length patterns of the human knee joint, *J. Biomechanics*, 12, 727, 1979.

42. Dortmans, L., Jans, H., Sauren, A., and Huson, A., Nonlinear dynamic behavior of the human knee joint I. Postmortem frequency domain analyses, *ASME J. Biomechanical Eng.*, 113, 387, 1991.

43. Dortmans, L., Jans, H., Sauren, A., and Huson, A., Nonlinear dynamic behavior of the human knee joint II. Time-domain analyses: effects of structural damage in postmortem experiments, *ASME J. Biomechanical Eng.*, 113, 392, 1991.

44. Dufek, J.S. and Bates, B.T., The evaluation and prediction of the impact forces during landings, *Med. Sci. Sports Exercise*, 22, 370, 1990.

45. Edwards, R.G., Lafferty, J.F., and Lange, K.O., Ligament strain in the human knee joint, *J. Basic Eng.*, 92, 131, 1970.

46. Engin, A.E. and Akkas, N., Application of a fluid-filled spherical sandwich shell as a biodynamic head injury model for primates, *Aviation Space Environ. Med.*, 49, 120, 1978.

47. Engin, A.E. and Moeinzadeh, M.H., Dynamic modelling of human articulating joints, *Math. Modelling*, 4, 117, 1983.

48. Engin, A.E. and Tumer, S.T., An innovative approach to the solution of highly nonlinear dynamics problems associated with joint biomechanics, Proceedings of the 1991 Biomechanics Symposium, Ohio State University, Columbus, OH, June 1991, 225.

49. Engin, A.E. and Tumer, S.T., Improved dynamic model of the human knee joint and its response to impact loading on the lower leg, *ASME J. Biomechanical Eng.*, 115, 137, 1993.

50. Essinger, J.R., Leyvraz, P.F., Heegard, J.H., and Robertson, D.D., A mathematical model for the evaluation of the behavior during flexion of condylar-type knee prostheses, *J. Biomechanics*, 22, 1229, 1989.

51. Figgie, H.E., Bahniuk, E.H., Heiple, K.G., and Davy, D.T., The effects of tibial-femoral angle on the failure mechanics of the canine anterior cruciate ligament, *J. Biomechanics*, 19, 89, 1986.

52. France, E.P., Daniels, A.U., Globe, E.M., and Dunn, H.K., Simultaneous quantitation of knee ligament forces, *J. Biomechanics*, 16, 553, 1983.

53. Gear, C.W. and Petzold, L.R., ODE methods for the solution of differential/algebraic systems, *SIAM J. Numer. Anal.*, 21, 716, 1984.

54. Gear, C.W., Leimkuhler, B., and Gupta, G.K., Automatic integration of Euler-Lagrange equations with constraints, *J. Comp. Appl. Math.*, 12, 77, 1985.

55. Gear, C.W. and Petzold, L.R., Differential-algebraic equation index transformations, *SIAM J. Sci. Stat. Comp.*, 9, 39, 1988.

56. Gerald, C.F. and Wheatley, P.O., *Applied Numerical Analysis*, Addison-Wesley, Reading, 1989, 132.

57. Gill, H.S. and O'Connor, J.J., Biarticulating two-dimensional computer model of the human patello-femoral joint, *Clin. Biomechanics*, 11, 81, 1996.

58. Girgis, F.G., Marshall, J.L., and Al Monajem, A.R.S., The cruciate ligaments of the knee joint, *Clin. Orthop.*, 106, 216, 1975.

59. Glos, D.L., Butler, D.L., Groos, E.S., and Levy, M.S., *In vitro* evaluation of an implantable force transducer (IFT) in a patellar tendon model, *ASME J. Biomechanical Eng.*, 115, 335, 1993.

60. Grood, E.S. and Hefzy, M.S., An analytical technique for modeling knee joint stiffness I. Ligamentous forces, *ASME J. Biomechanical Eng.*, 104, 330, 1982.

61. Grood, E.S. and Suntay, W.J., A joint coordinate system for the clinical description of three-dimensional motions: application to the knee, *ASME J. Biomechanical Eng.*, 105, 136, 1983.

62. Grood, E.S., Hefzy, M.S., and Lindenfield, T.N., Factors affecting the region of most isometric femoral attachments I. The posterior cruciate ligament, *Am. J. Sports Med.*, 17, 197, 1989.

63. Hatze, H., The meaning of the term *biomechanics*, *J. Biomechanics*, 7, 189, 1974.

64. Heegaard, J., Leyvraz, P.F., Curnier, A., Rakotomana, L., and Huiskes, R., Biomechanics of the human patella during passive knee flexion, *J. Biomechanics*, 28, 1265, 1995.

65. Hefzy, M.S. and Grood, E.S., Sensitivity of insertion locations on length patterns of anterior cruciate ligament fibers, *ASME J. Biomechanical Eng.*, 108, 73, 1986.

66. Hefzy, M.S. and Grood, E.S., Review of knee models, *Appl. Mechanics Rev.*, 41, 1, 1988.

67. Hefzy, M.S., Jackson, W.T., Saddemi, S.R., and Hsieh, Y.F., Effects of tibial rotations on patellar tracking and patello-femoral contact area, *J. Biomed. Eng.*, 14, 329, 1992.

68. Hefzy, M.S. and Yang, H., A three-dimensional anatomical model of the human patello-femoral joint to determine patello-femoral motions and contact characteristics, *J. Biomed. Eng.*, 15, 289, 1993.

69. Hafzy, M.D. and Abdel-Rahman, E., Dynamic modeling of the human knee joint: formuation and solution techniques, *Biomed. Eng. Appl. Basis Commun.*, 7, 5, 1995.

70. Hefzy, M.S. and Cooke, T.D.V., A review of knee models: 1996 update, *Appl. Mech. Rev.*, 49, S187, 1996.

71. Hindmarsh, A.C., ODEPACK, a systematized collection of ODE solvers, in *Scientific Computing*, Vol. 1, Stepleman, R.S. et al., Eds., North-Holland, Amsterdam, 1983, 55.

72. Hirokawa, S., Three-dimensional mathematical model analysis of the patello-femoral joint, *J. Biomechanics*, 24, 659, 1991.

73. Hof, A.L., Pronk, C.N.A., and van Best, J.A., Comparison between EMG to force processing and kinetical analysis for the calf muscle moment in walking and stepping, *J. Biomechanics*, 20, 167, 1987.

74. Hughston, J.C., Bowden, J.A., Andrews, J.R., and Norwood, L.A., Acute tears of the posterior cruciate ligament, *J. Bone Joint Surg.*, 62-A, 438, 1980.

75. Huiskes, R. and Blankevoort, L., The relationship between knee joint motion and articular surface geometry, in *Biomechanics of Diarthrodial Joints*, Vol. 2, Springer-Verlag, New York, 1990, 269.

76. Huson, A., Biomechanische Probleme des Kniegelenks, *Orthopaede*, 3, 119, 1974.

77. Jackson, K.R. and Sacks-Davis, R., An alternative implementation of variable step-size multistep formulae for stiff ODEs, *ACM Trans. Math. Software*, 6, 295, 1980.

78. Jans, H.W.J., Dortmans, L.J.M.G., Sauren, A.A.H.J., and Huson, A., An experimental approach to evaluate the dynamic behavior of the human knee, *ASME J. Biomechanical Eng.*, 110, 69, 1988.

79. van Kampen, A., The three-dimesional tracking pattern of the patella, Ph.D. dissertation, University of Nijmegen, The Netherlands, 1987.

80. Kao, R., A comparison of Newton-Raphson methods and incremental procedures for geometrically nonlinear analysis, *Comp. Struc.*, 4, 1091, 1974.

81. Kaufman, K.R., Muscle and joint forces in the knee during isokinetic exercise, *J. Biomechanics*, 23, 711, 1990.

82. LaFortune, M.A., Cavanagh, P.R., Summer, H.J., and Kalenak, A., Three-dimensional kinematics of the knee during walking, *J. Biomechanics*, 25, 347, 1992.

83. Lindbeck, L., Impulse and moment of impulse in the leg joints by impact from kicking, *ASME J. Biomechanical Eng.*, 105, 108, 1983.

84. Ling, Z.-K., Guo, H.-Q., and Boersma, S., Analytical study on the kinematic and dynamic behaviors of a knee joint, *Med. Eng. Phys.*, 19, 29, 1997.

85. Loos, W.C., Fox, J.M., Blazina, M.E., Del Pizzo, W., and Friedman, M.J., Acute posterior cruciate ligament injuries, *Am. J. Sports Med.*, 9, 86, 1981.

86. MacKinnon, C.D. and Winter, D.A., Control of the whole body balance in the frontal plane during human walking, *J. Biomechanics*, 26, 633, 1993.

87. Macquet, P.G., Van De Berg, A.J., and Simonet, J.C., Femorotibial weight-bearing areas: experimental determination, *JBJS*, 57-A, 766, 1975.

88. Menschik, A., Mechanik des Kniegelenkes, Teil 1, *Z. Orthoped.*, 112, 481, 1974.

89. Menschik, A., Mechanik des Kniegelenkes, Teil 2, *Z. Orthoped.*, 113, 388, 1975.

90. Menschik, A., The basic kinematic principle of the collateral ligaments, demonstrated on the knee joint, in *Injuries of the Ligaments and Their Repair*, Chapchal, G., Ed., Thieme, Stuttgart, 1977, 9.

91. Mills, O.S. and Hull, M.L., Apparatus to obtain rotational flexibility of the human knee under moment loads *in vivo*, *J. Biomechanics*, 24, 351, 1991.

92. Mills, O.S. and Hull, M.L., Rotational flexibility of the human knee due to various varus/valgus and axial moments *in vivo*, *J. Biomechanics*, 24, 673, 1991.

93. Moeinzadeh, M.H., Two- and three-dimensional dynamic modeling of the human joint structures with special application to the knee joint, Ph.D. dissertation, Ohio State University, Columbus, OH, 1981.

94. Moeinzadeh, M.H., Engin, A.E., and Akkas, N., Two-dimensional dynamic modeling of the human knee joint, *J. Biomechanics*, 16, 253, 1983.

95. Moeinzadeh, M.H. and Engin, A.E., Response of a two-dimensional dynamic model to externally applied forces and moments, *J. Biomedical Eng.*, 105, 281, 1983.

96. Moeinzadeh, M.H. and Engin, A.E., Dynamic modeling of the human knee joint, in *Computational Methods in Bioengineering*, Vol. 9, American Society of Mechanical Engineering, Chicago, 1988, 145.

97. Moffat, C.A., Harris, E.H., and Haslam, E.T., An experimental and analytical study of the dynamic properties of the human leg, *J. Biomechanics*, 2, 373, 1969.

98. Morrison, J.B., The mechanics of the knee joint in relation to normal walking, *J. Biomechanics*, 3, 51, 1970.

99. Mow, V.C., Ateshian, G.A., and Spilker, R.L., Biomechanics of diarthrodial joints: a review of twenty years of progress, *ASME J. Biomechanical Eng.*, 115, 460, 1993.

100. Muller, W., *The Knee: Form, Function and Ligament Reconstruction*, Springer-Verlag, Berlin, 1982.

101. Muller, W., Kinematics of the cruciate ligaments, in *The Cruciate Ligaments*, Feagin, J.A., Jr., Ed., Churchill Livingstone, New York, 1988.

102. Mungiole, M. and Martin, P.E., Estimating segment inertial properties: comparison of magnetic resonance imaging with existing methods, *J. Biomechanics*, 23, 1039, 1990.

103. Nigg, B.M., Cole, G.K., and Nachbauer, W., Effects of arch height of the foot on angular motion of the lower extremities in running, *J. Biomechanics*, 26, 909, 1993.

104. Ogata, K., McCarthy, J.A., Dunlap, J., and Manske, P.R., Pathomechanics of posterior sag of the tibia in posterior cruciate deficient knees: an experimental study, *Am. J. Sports Med.*, 16, 630, 1988.

105. Petzold, L.R., Differential/algebraic equations are not ODEs, *SIAM J. Sci. Stat. Comput.*, 3, 367, 1982.

106. Petzold, L.R., A description of DASSL: a differential/algebraic system solver, in *Scientific Computing*, Vol. 1, Stepleman, R.S. et al., Eds., North-Holland, Amsterdam, 1983, 65.

107. Platt, D., Wilson, A.M., Timbs, A., Wright, I.M., and Goodship, A.E., Novel force transducer for the measurement of tendon force *in vivo*, *J. Biomechanics*, 27, 1489, 1994.

108. Pope, M.H., Crowninshield, R., Miller, R., and Johnson, R., The static and dynamic behavior of the human knee *in vivo*, *J. Biomechanics*, 9, 449, 1976.

109. Race, A. and Amis, A.A., The mechanical properties of the two bundles of the human posterior cruciate ligament, *J. Biomechanics*, 27, 13, 1994.

110. Radin, E.L. and Paul, I.L., A consolidated concept of joint lubrication, *J. Bone Joint Surg.*, 54-A, 607, 1972.

110a. Rahman, E.M. and Hefzy, M.S., Three-dimensional dynamic behaviour of the human knee joint under impact loading, *JJBE: Med. Eng. Phys.*, 20, 4, 1998.

111. Rogers, G.J., Milthorpe, B.K., Muratore, A., and Schindhelm, K., Measurement of mechanical properties of the anterior cruciate ligament, *Aust. J. Phys. Eng.*, 8, 168, 1985.

112. Rohrle, H., Scholten, R., Sigolotto, C., Solbach, W., and Kellner, H., Joint forces in the human pelvis-leg skeleton during walking, *J. Biomechanics*, 17, 409, 1984.

113. Seedhom, B.B., Dowson, D., and Wright, V., The load-bearing function of the menisci: preliminary study, in *The Knee Joint, Proceedings of the International Congress*, Excerpta Medica, Rotterdam, 1974, 37.

114. Seireg, A. and Arvikar, R.J., The prediction of muscular load sharing and joint forces in the lower extremities during walking, *J. Biomechanics*, 8, 89, 1975.

115. van Soest, A.J., Schwab, A.L., Bobbert, M.F., and van Ingen-Schenau, G.J., The influence of the biarticularity of the gastrocnemius muscle on vertical jumping achievement, *J. Biomechanics*, 26, 1, 1993.

116. Soslowsky, L.J., Flatow, E.L., Bigliani, L.U., and Mow, V.C., Articular geometry of the glenohumeral joint, *Clin. Orthopedics Rel. Res.*, 285, 181, 1992.

117. Takai, S., Woo, S.L.-Y., Livesay, G.A., Adams, D.J., and Fu, F.H., Determination of the *in situ* loads on the human anterior cruciate ligament, *J. Orthopaed. Res.*, 11, 686, 1993.

118. Tumer, S.T. and Engin, A.E., Three-body segment dynamic model of the human knee, *ASME J. Biomechanical Eng.*, 115, 350, 1993.

119. Tumer, S.T., Wang, I., and Akkas, N., A planar dynamic anatomic model of the human lower limb, *Biomed. Eng. Appl. Basis Commun.*, 7, 365, 1995.

120. Vaughan, C.L., Davis, B.L., and O'Connor, J.C., *Dynamics of Human Gait*, Human Kinetics, Champaign, IL, 1992, 20.

121. Wahrenberg, H., Lindbeck, L., and Ekholm, J., Dynamic load in the human knee joint during voluntary active impact to the lower leg, *Scand. J. Rehabil. Med.*, 10, 93, 1978.

122. Wahrenberg, H., Lindbeck, L., and Ekholm, J., Knee muscular moment, tendon tension force and EMG during vigorous movement in man, *Scand. J. Rehabil. Med.*, 10, 99, 1978.

123. Walker, P.S. and Hajek, J.V., The load bearing area in the knee joint, *J. Biomechanics*, 5, 581, 1972.

124. Walker, P.S., Kurosawa, M.D., Rovick, M.A., and Zimmerman, B.S., External knee joint design based on normal motion, *J. Rehabil. Res. Dev.*, 22, 9, 1985.

125. Walker, P.S. and Garg, A., Range of motion in total knee arthroplasty: a computer analysis, *Clin. Orthopedics Rel. Res.*, 262, 227, 1991.

126. Wang, C.J., Walker, P.S., and Wolf, B., The effects of flexion and rotation on the length patterns of the ligaments of the knee, *J. Biomechanics*, 6, 587, 1973.

127. Wang, C.J. and Walker, P.S., Rotatory laxity of the human knee joint, *J. Bone Joint Surg.*, 56-A, 161, 1974.

128. Warren, L.F., Marshall, J.L., and Girgis, F., The prime static stabilizer of the medial side of the knee, *J. Bone Joint Surg.*, 56-A, 665, 1974.

129. Wismans, J., A three-dimensional mathematical model of the human knee joint, Ph.D. dissertation, Eindhoven University of Technology, Eindhoven, The Netherlands, 1980.

130. Wismans, J., Veldpaus, F., Janssen, J., Huson, A., and Strulen, P., A three-dimensional mathematical model of the knee joint, *J. Biomechanics*, 13, 677, 1980.

131. Wongchaisuwat, C., Hemami, H., and Buchner, H.J., Control of sliding and rolling at natural joints, *ASME J. Biomechanical Eng.*, 106, 368, 1984.

132. Woo, S. L.-Y., Hollis, M., Roux, R.D., Gomez, M.A., Inoue, M., Kleiner, J.B., and Akeson, W.H., Effects of knee flexion on the structural properties of the rabbit femur anterior cruciate ligament-tibia complex (FATC), *J. Biomechanics*, 20, 557, 1987.

133. Woo, S. L.-Y., Hollis, M., Adams, D.J., Lyon, R.M., and Takai, S., Tensile properties of the human femur anterior cruciate ligament-tibia complex: the effects of specimen age and orientation, *Am. J. Sports Med.*, 19, 217, 1991.

134. Woo, S. L.-Y., Johnson, G.A., and Smith, B.A., Mathematical modeling of ligaments and tendons, *ASME J. Biomechanical Eng.*, 115, 468, 1993.

135. Yasuka, K., Erickson, A.R., Johnson, R.J., and Pope, M.H., Dynamic strain behavior in the medial collateral and anterior cruciate ligaments during lateral impact loading. *Trans. Orthopaed. Res. Soc.*, 17, 127, 1992.

136. Zoghi, M., Hefzy, M.S., Jackson, W.T., and Fu, K.C., A three-dimensional morphometrical study of the distal human femur, *J. Eng. Med.*, 206, 147, 1992.

137. Zoghi, M., Solving the 3-dimensional contact problem between femur and menisci and tibia, Doctoral dissertation, University of Toledo, Toledo, OH, 1997.

2

Techniques and Applications of Adaptive Bone Remodeling Concepts

Nicole M. Grosland
University of Iowa

Vijay K. Goel
University of Iowa

Roderic S. Lakes
University of Iowa

2.1 Introduction and Synopsis

Despite the apparent inanimate nature of bone, bone is a dynamic living material constantly being renewed and reconstructed throughout the lifetime of an individual. Bone deposition and bone resorption typically occur concurrently, so that bone is remodeled continually. It is this adaptive remodeling process, driven partially in response to functional requirements, that distinguishes living structural materials from other structural solids. As a complex biological phenomenon, adaptive bone remodeling has played a dominant role in the study of bone physiology and biomechanics for over a century, and has been active biologically for as long as there have been vertebrates.

The relationship between the mass and form of a bone to the forces applied to it was appreciated by Galileo,[1] who is credited with being the first to understand the balance of forces in beam bending and applying this understanding to the mechanical analysis of bone. Julius Wolff[2] published his seminal 1892 monograph on bone remodeling; the observation that bone is reshaped in response to the forces acting on it is presently referred to as Wolff's law. Many relevant observations regarding the phenomenology of bone remodeling have been compiled and analyzed by Frost.[3,4] Despite the general acceptance that mechanical stimulation influences bone homeostasis and adaptation, controversy remains as to the governing mechanical stimuli: how mechanical signals are transduced at the cellular and subcellular levels, and whether electrical and molecular phenomena coincident with mechanical stimulation mediate cellular responses.[5]

The development of a theoretical framework for the prediction of bone remodeling and the clinical implementation of this framework are of particular interest. An all-inclusive understanding of bone remodeling has the following potential in clinical practice: the reduction, treatment, or possible prevention of osteoporotic bone loss; acceleration of fracture healing; and the optimization of implant design. In search of this goal, mechanistic and phenomenological theories of bone remodeling have been proposed.[6-12]

The following sections provide an overview of the methods hypothesized to predict the phenomenon of adaptive bone remodeling. Following a brief review of bone morphology, special emphasis is placed on adaptive remodeling: theoretical and experimental investigations, proposed theoretical models of bone adaptation, and the possible causal mechanisms responsible for the adaptive bone remodeling processes.

2.2   Structure of Bone

The development of bone from embryonic to adult size depends on the orderly processes of mitotic divisions, cellular growth, and structural remodeling. Bone has been recognized as a highly complex system, a multifunctional tissue subjected to a large number of interrelated biochemical, biophysical, and biological processes.[13] In turn, mechanical and geometrical properties of bone have been attributed to these processes.

The sizes, shapes, and structures of human skeletal bones are quite well known. Each bone possesses a characteristic pattern of ossification and growth, a characteristic shape, and features that indicate its functional relationship to other bones, muscles, and to the body structure as a whole. The shape and surface features of each bone are related to its functional role in the skeleton. Long bones, for example, function as levers during body movement. Bones that support the body are massive, with large articulating surfaces and processes for muscular attachment. Because the primary responsibility of the skeleton is structural, bone has acquired the unfortunate reputation of being a simple material.[14]

Histological Organization

The following is a brief discussion with regard to the basic histological organization of bone, for it is with this understanding that the significance of the structure may be assessed. Bone represents a complex, highly organized, connective tissue, characterized physically by its hardness, rigidity, and strength, and microscopically by relatively few cells and considerable intercellular substance, formed of mineralized fibers and cement. It has a rich vascular supply and is the site of considerable metabolic activity. At the lowest level, bone may be categorized as a composite material composed of a fibrous protein, collagen, stiffened by an extremely dense filling of inorganic calcium phosphate (hydroxyapatite). Bone has additional constituents, namely, water and some ill-understood amorphous polysaccharides and proteins which accompany living cells and blood vessels.

Bone Cells

Four types of bone cells are commonly recognized: *osteoblasts, osteocytes, osteoclasts,* and *bone lining cells*.[15] Osteoblasts, osteocytes, and the bone lining cells arise from primitive mesenchymal cells, called osteoprogenitor cells, within the investing connective tissue. Bone formation is carried out by active osteoblasts, which synthesize and secrete the proteins and other organic components of the bone matrix.[16] This process creates an organic matrix known as osteoid, within which calcium and phosphate are subsequently deposited in amorphous masses. The inorganic or mineral phase constitutes approximately 50% of bone by volume and is composed of calcium crystals primarily in the form of hydroxyapatite.[17] Hydroxyapatite crystals precipitate in an orderly fashion around collagen fibers present in the osteoid. The osteoid rapidly calcifies (approximately 70% calcification after a few days), reaching maximal calcification within several months.[18]

As the completely mineralized bone accumulates and surrounds the osteoblast, that cell loses its synthetic activity and becomes an interior osteocyte. Although encased in the mineralized bone matrix, osteocytes maintain contact with other osteocytes, osteoblasts, and bone lining cells via an extensive network of small, fluid-containing canals, or canaliculi. The bone lining cells are resting cells located on inactive bone surfaces which represent more than 80% of the trabecular and endocortical surfaces of adult bone.[15,19] These cells represent a terminal differentiation of the osteoblasts and are thinly extended over the bone surface. In general, mitotic activity is absent. Upon stimulation, however, the bone lining cells may be activated to form a layer of osteoblasts. Osteoclasts, on the other hand, are multinucleated giant cells with the capability of removing bone tissue in a process referred to as osteoclasis or bone resorption.

Bone Tissue

At the macroscopic level, adult bone tissue is broadly divided into two distinguishable forms: cortical bone, also referred to as *compact* bone, and trabecular bone, also referred to as *spongy* or *cancellous* bone (Fig. 2.1). Trabecular and cortical bone differ in histological structure, gross appearance, location, and function. Dense cortical bone comprises the diaphysis of appendicular long bones while a thin shell encompasses the metaphysis. Cancellous, or trabecular, bone exists as a three-dimensional, interconnected network of rods and plates which delimit a labyrinthine system of intercommunicating spaces that are occupied by bone marrow. This porous, highly vascular tissue reduces the weight of the bone, while providing space for bone marrow where blood cells are produced.

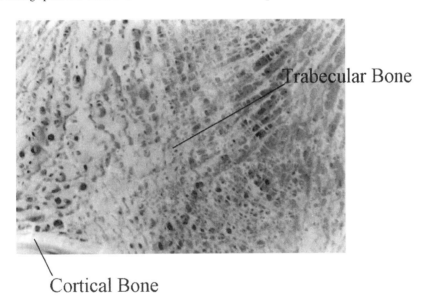

FIGURE 2.1   Lamellar organization — appearance and distribution of trabecular and cortical bone.

Although it constitutes only 20% of the skeleton, trabecular bone has a greater overall surface area than does cortical bone and is considered to possess greater metabolic activity. Relative density (i.e., the ratio of specimen density to that of fully dense cortical bone — usually 1.8 g/cc) provides the criterion upon which the classification of bone tissue as cortical or cancellous is based. The relative density of trabecular bone varies from 0.05 to about 0.7 while that of cortical bone is approximately 0.7 to about 0.95.[20] The external surface of bone is covered by a periosteum consisting of a fibrous connective tissue outer layer and a cellular inner layer. The periosteum not only serves for the attachment of muscles, but aids in protection and provides additional strength to the bone. Moreover, the periosteum provides a route for circulatory and nervous supply, while actively participating in bone growth and repair.[16]

While all bone tissue in mammals contains cells, fibers, and cement, their relative amounts and arrangements vary. Because the chemical, molecular, and cellular components are similar among bone types, the variability in properties of bone has been attributed to the differences in the organization of these elements. In general, bone microstructure can be divided into three broad categories: (1) woven bone; (2) primary bone (primary lamellar, plexiform, primary osteons); and (3) secondary bone.

Woven bone is architecturally arranged as a close network of fine trabeculae. It is nonlamellar and generally less dense than other types of bone. It should be noted that the reduction in density is a function of the loose packing of collagen fibers and large porosities rather than reduced mineralization. The collagen in woven bone has fine fibers, approximating 0.1 μm in diameter, and oriented almost randomly. Consequently, it is difficult to make out any preferred direction over distances in excess of a few micrometers (μm). Typically, woven bone proliferates rapidly, most notably in the fetus and during callus formation in fracture repair. Equally rapid woven bone formation can result from damage to, or tension on, the periosteum.

In contrast to woven bone, primary bone requires a pre-existing substrate for deposition. Consequently, lost trabeculae may not be replaced, unless done so by woven bone, and then remodeled. Furthermore, primary bone is divided into three morphologically distinct categories: primary lamellar, plexiform, and primary osteons.

Lamellar bone is distinguished histologically by its multilayered structure. Primary lamellar bone is arranged circumferentially around the endosteal and periosteal surfaces of whole bones. Primary lamellar bone can become increasingly dense. Compact lamellar bone superficially resembles plywood in section, as if numbers of thin plates were cemented together. A series of concentric plates characterizes the cross-sectional appearance.[21] Cancellous bone is also composed of primary lamellae, with a large surface area intimately contacting marrow. In general, primary lamellar bone exhibits superior mechanical strength.

Like woven bone, plexiform bone is deposited rapidly, but exhibits mechanical qualities superior to those of woven bone. Analogous to primary lamellar bone, plexiform must be deposited on pre-existing surfaces. Structurally, however, plexiform bone resembles highly oriented cancellous bone. Plexiform is predominatly seen in larger, rapidly growing animals such as young cows, and has been observed in growing children.[22]

When the accretion of lamellar bone surrounds a centrally placed blood vessel, a concentrically arranged osseous structure is created. It is this architecture that distinguishes the Haversian system. A set of concentric lamellae (from a few to as many as 20), in conjunction with associated bone cells (osteocytes) and a central vascular channel, constitutes an osteon. Osteons are elongated, almost solid cylinders largely directed in the long axis of the bone.

Osteons are categorized as primary or secondary. Primary osteons are the first to be laid down in early life. Note that primary bone is new bone deposited in a space where bone failed to exist previously, although it may be fabricated on an existing bone surface. By contrast, secondary bone is the product of the resorption of pre-existing bone tissue and the deposition of new bone in its place. In cortical bone, the result of osteoclastic resorption and subsequent osteoblastic formation is a secondary osteon. Consequently, secondary osteons replace primary osteons and are subject to continual resorption and renewal throughout life. The process is referred to as internal remodeling. Primary osteons are relatively small; they have no cement lines; there are no fragments or wedges of interstitial bone between them; and the central canal may contain two or three vessels. In contrast, secondary osteons are generally larger structures. They are surrounded by narrow cement lines and between them reside irregular pieces of lamellar bone (interstitial bone), many of which are remnants of former osteons removed during remodeling.[21] The cement lines bounding secondary osteons tend to be irregular and represent lines of reversal, indicating the change from bone resorption to deposition. The absence of these lines in primary osteons establishes the prominent morphological distinction from secondary osteons. After initial deposition, all types of bone are subject to secondary reconstruction or remodeling.

2.3   Adaptive Remodeling of Bone

Phenomena

Living bone is continually undergoing processes, collectively termed *remodeling*, of deposition and resorption, partially but not totally driven by changes in its mechanical load environment.[15] This dynamic aspect of bone tissue has the effect of providing strength in direct response to weightbearing stress. Adaptive remodeling may be conveniently recognized as external and internal, although the cellular mechanisms are the same for both and the processes overlap in time.[21] External remodeling is concerned with the architecture of bones (i.e., geometry and form), while internal processes alter the bone structure histologically. Remodeling may replace the matrix material while leaving the bone as a whole unchanged, or may produce alterations in the shape, internal architecture, or mineral content of the bone. Although the general shapes of bones are established genetically, other forces are at play. The actions of muscles, in addition to their places of origin and insertion, introduce important mechanical factors that influence the external shape and internal arrangement of trabeculae. Erect posture, for example, in combination with gravity, greatly influences the internal architecture of the vertebrae of the axial skeleton (Fig. 2.2) and long bones of the appendicular skeleton.

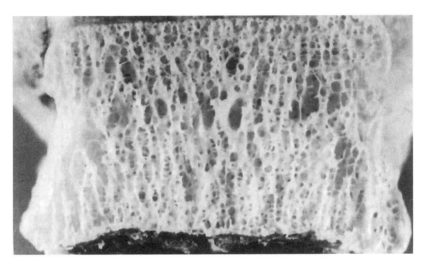

FIGURE 2.2   Longitudinal section through a vertebral body illustrating the trabecular network. *Source*: White and Panjabi, *Clinical Biomechanics of the Spine*, J.B. Lippincott Co., 1990, p. 41. With permission.

General Description and Clinical Observations

In a healthy adult who maintains a consistent level of physical activity, a balance between bone resorption and formation exists so that there is no net change in bone mass. In general, adaptive bone remodeling is acknowledged as error-driven. That is, mechanical loads upon bone must deviate from normal values by a sufficient amount (error signal) to initiate a remodeling response. If the threshold stress is not exceeded, no remodeling response occurs. In addition, saturation limits beyond which bone refrains from adapting are assumed by most theories. The mechanisms for regulation of the remodeling process are largely unknown, but undoubtedly involve local regulatory factors, a combination of physical factors as a result of weightbearing stress, and the effects of calcium-regulating hormones impinging on the different bone cell populations.[22,23] The turnover rate of bone is undeniably high. In a young adult, approximately one-fifth of the skeleton is resorbed and then rebuilt or replaced annually.[16] It should be noted that regional as well as local differences exist in the rate of turnover. As a result, not every part of every bone will be equally affected.

   Although the mechanosensory system in bone tissue has not been identified, it appears reasonable to assume that, when living bone is deformed, the mechanical strain signal is transduced to the bone cell population.[15] Mechanical loading has been cited as an important factor shifting the remodeling balance in favor of bone formation in adults. Mechanical factors are responsible for adjusting the strength of bone in response to the demands placed upon it. The greater the physical stress to which the bone is subjected, the greater the rate of bone deposition. On the other hand, loss of bone mass occurs in response to the removal of mechanical stress, as in persons who undergo prolonged bed confinement or those in prolonged space flight. Such changes in response to altered mechanical loading conditions have significant clinical implications.

At an estimated rate of 600,000 operations yearly, artificial joint replacements constitute one of the major surgical advances of this century, second only to dental reconstruction as an invasive treatment of bodily ailments.[24] Adaptive bone remodeling has been cited as an important factor affecting the long-term post-operative behavior of joint replacements. Clinical, as well as experimental, studies have demonstrated that factors such as prosthesis position, patient activity level, body weight, fixation technique, and component material properties significantly affect the success rate of total joint arthroplasty.[25-27] These factors, whether in conjunction or individually, contribute to the nature of the mechanical environment at the bone-implant interface. The long-term success of any implant is going to be dependent on the biomechanical as well as biochemical compatibility of the implant.[28] For example, a decrease in stress in the proximal femur as a result of load transmission through the implant stem, past the proximal femur, to the midshaft is thought to be a possible explanation for the resorption in the calcar region and subsequent loosening of hip implants.[29] The changes in bone structure following prosthetic joint replacement have been studied extensively, especially for total hip replacements,[30,31] and total knee replacements.[32,33] Advanced stress analysis methods, such as finite element (FE) modeling, have proven to contribute significantly to such research endeavors.

Quantitative Experiments

While clinical inquiries provide the ultimate evaluation of long-term implant-induced remodeling, the cost, duration, and ethical considerations involving human experimentation prolong feedback to the implant designers. The advent of modern computing capabilities, in conjunction with numerical stress analysis techniques, enabled researchers to relate bone mechanics to the observed bone structure. The aforementioned mathematical descriptions have enabled bone modeling and remodeling simulations to be implemented in combination with the finite element method (FEM).[10,11,34-38] Most of these evaluations have sought to procure a characteristic mechanical stimulus, or collection of stimuli which predict most realistically, adaptive remodeling in response to distinct loading modalities.

As previously mentioned, it is common for bone remodeling theories to be coupled with the finite element method. In general, such simulations initiate with a given model geometry, initial density distribution, and a set of selected applied load cases. The remodeling equations are employed to update the internal density distribution and/or external geometry incrementally. The model is considered to have converged once the change in density and/or geometry with each increment is small. Validation studies reveal that these computer simulations enable accurate predictions of long-term formation and resorption of bone around orthopedic implants in animals and in humans. Consequently, the incentive for continued investigations aimed at establishing the specific factors governing the adaptation response of bone is great. To date, the majority of work in this area has focused on the femur, knee, and more recently the spine.

The validity of such finite element models must be assessed by experimental verification. Brown et al.[39] made an attempt to correlate the Rubin-Lanyon turkey ulna model with finite element modeling to establish the mechanical parameters associated with bone remodeling. Functionally isolated turkey ulnae were selected, enabling the loading conditions to be characterized completely while the periosteal adaptive responses were monitored and quantified after four and eight weeks of loading. Subsequently, their three-dimensional FE model of the ulna was validated against a normal strain-gauged turkey ulna under identical loading conditions. Twenty-four mechanical parameters were compared in an attempt to correlate the FE results with those obtained experimentally. The pattern of periosteal bone remodeling was most highly correlated with strain energy density and longitudinal shear stress. Recently, Adams[5] extended the preliminary work of Brown et al.[39] to 43 candidate mechanical parameters, further investigating whether the initiation of appositional bone formation in the turkey ulna can be predicted by examining changes in mechanical factors associated with controlled loading regimens.

Beaupré et al.[40] combined adaptive remodeling techniques and finite element analyses to forecast the evolution of density in the human proximal femur. A two-dimensional finite element model of the human femur was subjected to three loading conditions to establish the daily tissue stress level stimulus. Representative loads consisted of a single-legged stance and extreme cases of abduction and adduction with respective daily load histories of 6000, 2000, and 2000 cycles. Based on the daily load history, the simulation was used to predict the density evolution from an initial homogeneous state. Density distributions were established after various iterations (i.e., 1, 15, 30) for remodeling with and without the inclusion of a lazy zone (i.e., a certain threshold level in over- or underloading must be exceeded prior to an adaptive response). As the number of time increments exceeded 30, the differences between the two models became more pronounced. The model incorporating the lazy zone showed little change (elemental density changes < 0.02 g·cm^{-3}), while in the absence of a lazy zone, the model continued to change as far as 125 iterations, predicting much higher density gradients. The more realistic density gradients predicted by the lazy zone may warrant attribution to some physiologic counterpart to which it is related. Orr et al.[41,42] embarked on a similar investigation into the bone remodeling induced by a femoral surface prosthesis. The density changes induced by a metal cap, a metal cap and central peg, and an epiphyseal plate surface prostheses were computed. It was assumed that there was total bone ingrowth in the prosthetic device, rigidly bonding the bone and implant.

Huiskes et al.[37] coupled the finite element method with numerical formulations of adaptive bone remodeling to investigate the relation between stress shielding and bone resorption in the femoral cortex around intramedullary prostheses such as those used in total hip arthroplasty (THA). A generalized, simple model of intramedullary fixation was implemented.[43] The FE model consisted of a two-dimensional axisymmetric straight bone and stem. Results indicated that the amount of bone resorption is largely dependent upon the rigidity and bonding properties of the implant; these results are compatible with animal experimental data on similar intramedullary configurations reported in the literature.[44-47]

Huiskes et al.[48] developed a three-dimensional FE model of a finger joint system. FE analysis was carried out to investigate the stress patterns in the structure as a whole and to establish the influences of material and design alternatives on these patterns. A follow-up investigation[49] was aimed at evaluating the aforementioned stress patterns at a local rather than global level, enabling a more detailed comparison with bone adaptive behavior.

Levenston et al.[50] employed computer modeling techniques to examine stress-related bone changes in the peri-acetabular region. They simulated the distribution of bone density throughout the natural pelvis as well as changes in bone density following total hip arthroplasty. The post-surgical models analyzed simulated fully fixed and loose bone-implant interfaces. The geometrical nature of the finite element model was based on a two-dimensional slice through the pelvis, passing through the acetabulum, pubic symphysis, and sacroiliac joint.[50,51] Initiating from a homogeneous bone density distribution, an incremental, time-dependent technique was employed to simulate the bone density distribution of the pelvis.

The average daily loading history was approximated with loads from a number of different activities along with the assumed daily frequencies of each. The simulations progressed until a stable bone density or state of little net bone turnover was achieved. The authors simulated the distribution of bone density in the natural pelvis as well as changes in bone density following total hip arthroplasty (THA). When loads representing multiple activities were incorporated, the predicted bone density for the natural pelvis was in agreement with that of the actual bone density distribution (Fig. 2.3). In contrast, the simulation restricted to a single-limb stance did not generate bone density distribution deemed realistic. This supports the concept that diverse loading plays a dynamic role in the development and maintenance of normal pelvic bone morphology. Utilizing the density distribution predicted of the natural bone, the finite element models were modified to investigate two designs of noncemented, metal-backed acetabular cups. A number of morphologic changes were predicted by these simulations.

The fully ingrown spherical component induced extensive bone resorption medial and inferior to the acetabular dome and bone hypertrophy near the interior rim; the fully loose component induced a lower level of bone loss as well as bone hypertrophy, by comparison. Acetabular components with no ingrowth transferred loads in a more physiologic manner than their fully fixed counterparts. The authors concluded

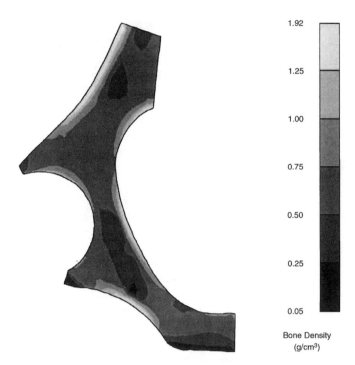

FIGURE 2.3   Predicted density distribution around the natural acetabulum subjected to a varied loading history. *Source*: Levenston, M.E. et al., *J. Arthroplasty*, 8, 595, 1993. With permission.

that an ideal state of complete bony fixation may yield unfavorable adaptive responses; hence, a successful acetabular component design must balance such considerations. It was interesting to note that the overall bone remodeling predicted around the acetabular components is much less destructive than that around the prosthetic femoral components.

A preliminary study by Goel and Seenivasan[52,53] applied a bone-adaptive remodeling theory to a basic ligamentous lumbar spine model. The change in shape of a two-motion segment model in response to axial compression and as a function of injury and stabilization was of primary interest. The vertebral bodies and discs were assumed to be cylindrical and have flat endplates. The simplified cylindrical shape was adopted in the attempt to validate the hypothesis that the bone adaptive remodeling applications yield the actual vertebral configuration. In response to an axially compressive load, the shapes of the remodeled vertebrae closely resembled the shape of an actual vertebral body (Fig. 2.4). The changes in shape observed in response to the fixation device were representative of stress shielding, characteristic of rigid fixation. Although the study demonstrated the feasibility of quantifying changes observed in the spinal segments following surgery, the simplicity of the model entailed limitations. Because spinal structures are inherently complex, the FE model utilized required considerable refinement.

In a follow-up study,[54,55] similar trends were observed in a more detailed model of the spine. An insignificant change in external geometry was not surprising because the models were derived directly from CT scans. As a result, few alterations were necessary. The opposite held true for internal remodeling, however. The internal remodeling algorithm converged for all governing loads, excluding torsion. The total strain energy density (TSED) for cancellous bone decreased as a function of iteration (or time), reaching a minimum at iteration 30 during compression. TSED is defined in the "Strain Energy Density (SED) Theory of Adaptive Bone Remodeling" subsection of the "Empirical Models" section of this chapter.

The elastic modulus distribution in the mid-transverse plane of the L4 vertebral body was higher in the postero- and central regions of the cross-section (Fig. 2.5). In the coronal view, elevated values were predicted centrally and adjacent to the endplates. An approximately uniform distribution of 15 MPa was

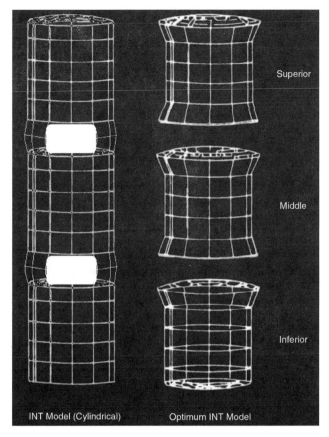

FIGURE 2.4 The optimal vertebral shape predicted via a cylindrical intact (INT) model in response to axial compression.

observed in other loading modes. Deviations from uniformity, however, were observed. For example, during extension, increased elastic moduli were predicted in the anterior and posterior regions of the transverse plane, the central region of the coronal plane, and the posterocentral and anterior inferior-most elements of the midsagittal plane. The distribution for flexion and lateral bending closely resembled that of extension, deviating slightly. The predicted inhomogeneous distribution following the remodeling procedures was in agreement with the experimental data.

In flexion, extension, and lateral bending modes, the cancellous bone region surrounds the neutral axis (bending axis). This is due to the load sharing role of posterior elements (ligaments in flexion, facets and ligaments in extension, and lateral bending modes). Thus, one would expect a smaller role of the stresses/strains in the bone remodeling process of the cancellous bone. Furthermore, in a healthy person, the muscular forces counteract the external bending moments and the ligamentous spine is only subjected to axial compression and small amounts of AP and lateral shear forces. Thus, the contributions of the bending moments toward the inhomogeneity of the cancellous bone should be minimal. The precise cause of nonconvergence in the axial torsional mode is not clear, but the ineffectiveness of torsional loads on the bone remodeling of a vertebral body is in agreement with similar predictions for the long bones.[56] This study has opened up a new research direction in the area of spinal biomechanics.

For example, the use of threaded interbody fusion cages for achieving spinal fusion has the potential to impart increased stability, while simultaneously reducing complications associated with the use of autogenous bone grafts. Interbody fusion devices are designed to facilitate stability as well as restore and maintain disc height. The BAK (BAK Interbody Fusion System, Spine-Tech) implant is a hollow, threaded cylinder accommodating multiple fenestrations to facilitate bone ingrowth and through-growth.

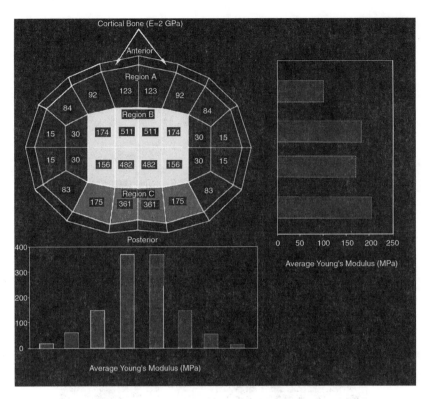

FIGURE 2.5   Predicted average elastic moduli distribution (in MPa, unless noted otherwise) in the transverse plane of the L4 vertebra subjected to 424.7 N axial compression.

Consequently, this device is an ideal candidate for exercising the aforementioned bone remodeling applications in conjunction with a finite element model of the spine.

Grosland et al.[57] are currently in the process of mimicking a spinal fusion, incorporating the BAK Interbody Fusion System in a refined finite element model of the spine (L3-L5). A bilateral anterior surgical approach was assumed. The internal remodeling algorithm utilized was based on a blend of various hypotheses reported in the literature.[37,38,58] A preliminary investigation has predicted a number of morphological alterations in response to the bone remodeling simulations following implantation of the device. During compression, the overlying and underlying bone directly adjacent to the device experienced bone hypertrophy (expressed as a percent change with respect to the intact model), while atrophy was induced laterally (see Fig. 2.6). Bone adjacent to the large holes of the device experienced minimal change. During flexion, extensive bone hypertrophy was induced anteriorly adjacent to the device, while atrophy was predicted posteriorly. The results clearly indicate that the vertebral bone, following cage implantation, undergoes both hypertrophy and atrophy as compared to the optimal intact bone density distribution. Bone growth in the anterior region predicted by the model is in agreement with experimental observations. Thus, the bone is likely to grow in and around the larger size holes of the BAK device, suggesting that in the long run the device will entrench itself into the denser bone.

Empirical Models

Following in the footsteps of Wolff, investigators began experimenting with mathematical descriptions of mechanical bone-mass regulation. Their theories provide a quantitative formulation of Wolff's law which states, qualitatively, that bone is an optimal structure relative to its mechanical requirements and possesses the ability to maintain an optimal configuration in response to a mechanical alteration. As stated originally, Wolff's law was neither quantitative nor mechanistic. The first quantitative demonstration that

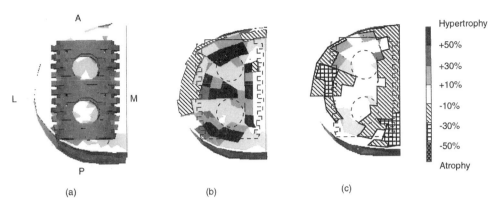

FIGURE 2.6 Interbody fusion system: (a) FE model of the BAK device positioned with respect to the inferior surface of L4 (A = anterior; P = posterior; L = lateral; and M = medial). Percent change in bone density adjacent to the BAK device with respect to the intact model during (b) 400 N compression and (c) 10 Nm flexion and 400 N preload. Dashed line indicates position of the device.

bone responds to its mechanical environment was presented by Koch,[59] although André[60] was the first to suggest that deformation rather than load could govern bone geometry. Martin[22] suggests that the idea may have been conceptualized and stated most clearly by D'arcy-Thompson:

> The origin, or causation, of the phenomenon would seem to lie partly in the tendency of growth to be accelerated under strain and partly in the automatic effects of shearing strain, by which it tends to displace parts which grow obliquely to the direct lines of tension and pressure, while leaving those in place which happen to lie parallel or perpendicular to those lines … accounting therefore for the rearrangement of … the trabeculae within the bone.[61]

Two types of bone mass changes may occur. Remodeling may affect the density of the bone and thereby its elastic moduli (internal remodeling) or its structural behavior (external remodeling). As a result of either remodeling process, the stresses and strains throughout the bone will be altered. That may in turn perpetuate a cascade necessitating further remodeling. The process continues until the remodeled bone density and shape are optimally suited to support the imposed loads. The precise nature of the feedback mechanism is neglected in the modeling of the adaptation process; it is only asserted that such a process exists. For example, numerous biological and biochemical constituents discussed previously are overlooked, or dealt with superficially.

Frost's Flexural Neutralization Theory

The flexural neutralization theory (FNT) of bone remodeling developed by Frost[7] in 1964 became the first mathematical formulation of bone remodeling as a function of mathematical variables. Frost suggested that changes observed in bone curvature, in combination with the polarity of tangential stress, are intimately associated with remodeling responses, namely, an increase in surface convexity favors bone resorption (osteoclastic activity), while bone deposition (osteoblastic activity) is promoted by a decrease in convexity. Initially, Frost theorized that there exists a *minimum effective stress* that must be exceeded to excite an adaptive remodeling response to mechanical overload.[7] In recent years, he has reformulated his theory in terms of strain rather than stress. Instead of speculating that strains below a certain threshold are "trivial" and evoke no adaptive response, Frost suggests that a *range* of strain values elicits no response.[89] Consequently, strains above this threshold evoke a positive adaptive response (i.e., deposition of bone), while those below the threshold induce a negative response (i.e., bone resorption).

The aforementioned FNT proposed by Frost, however, has been criticized on the basis that bones are naturally curved, and need be. It must be kept in mind that Frost's theory concerns load-induced changes in surface curvature rather than absolute curvature. Martin[22] suggests that if Frost originally expressed his theory in terms of a variable more directly related to strain and divorced from notions of local anatomic conformation, the confusion and debate may have been reduced. All controversy aside, Frost is commonly credited with providing the conceptual framework from which many of the current mechanical theories have been guided.

Pauwels' Stress Magnitude Theory

Pauwels[62] proposed a model for predicting the cortical thickness of diaphyseal bone as a function of the axial stresses due to bending. Accurate predictions were attained with respect to distortions in the cross-sectional geometry of a rachitic femur, through simplified assumptions relating surface remodeling to stress. Simplifying the initial femoral cross-section to a hollow elliptical geometrical configuration, the surface stress, σ_s, was calculated as a function of a simulated hip load. Alterations were made to the cortical thickness (T_c) via the following power function:

$$T_c = a + b\sigma_s^n \tag{2.1}$$

where, a, b, and n are arbitrary constants. A sequence of remodeling steps was established. Following an iterative process, the final stage (considered an equilibrium point) closely resembled the geometrical configuration of the actual bone. The rationale for the algorithm defining his model was not explained in detail; nonetheless, his efforts established the capabilities of a simplistic model to predict generalized adaptive geometries. Kummer[63] advanced a concrete form of Pauwels' theory which has been carefully reviewed by Firoozbakhsh and Cowin.[64] The cubic relationship between bone remodeling and stress is expressed as:

$$U = a[\sigma_s - \sigma_o)^2 (\sigma_i - \sigma_s) - (\sigma_i - \sigma_s)^3] \tag{2.2}$$

where, U is a measure of the bone remodeling (i.e., positive values indicate bone apposition, negative values indicate bone resorption); a is a proportionality factor related to the remodeling rate; σ_i is the actual stress; σ_s is the optimal equilibrium stress; and σ_o represents the lowest/highest tolerable bone stress. The cubic relationships developed by Kummer accounted for the adaptive changes associated with pressure necrosis, but neglected those associated with disuse atrophy.

Cowin's Adaptive Elasticity Theory

The mathematically rigorous and potentially powerful theory proposed by Cowin and colleagues,[56,65-67] was developed to describe the physiological adaptive behavior of bone. The basic hypothesis governing the thermomechanical continuum theory of adaptive elasticity is that the load-adapting properties of living bone can be modeled by a chemically reacting porous medium in which the rate of reaction is strain controlled. The objective was to model bone as a porous elastic solid and to model the normal adaptive processes that occur in bone remodeling as strain controlled mass deposition or resorption processes which modify the porosity of the porous elastic solid.[65] An implementation of this model[56] revealed that a nonhomogeneous cylindrical bone would become homogeneous when subjected to uniform stress. In addition, it was shown that remodeling will not occur in a long bone, such as the femur, as a result of a purely torsional load about its long axis.

   In the years that followed, Cowin and Firoozbakhsh[68] presented a somewhat less rigorous surface adaptation model in which bone assumed a site-specific homeostatic equilibrium strain state. Control equations, in which the rate of remodeling is proportional to the deviation from a reference (homeostatic) value were developed. Consequently, any aberrant strain state would influence bone remodeling in an attempt to reinstate homeostatic conditions via the following formula:

$$U = C_{ij}\left(e_{ij} - e_{ij}^o\right) \tag{2.3}$$

where U represents the rate of deposition or resorption; e_{ij} is the actual strain tensor, and e_{ij}^o is the homeostatic or reference strain tensor. The C_{ij} establishes a generalized matrix of remodeling coefficients. It should be noted that the authors relied on generality for the choice of C_{ij}, without reference to a biological basis. The values of the remodeling rate coefficients are necessary for a model to prove biologically useful, as the C_{ij} tensors contain coefficients for each component of strain. Experimental procedures indicate that the coefficients vary with each test model, consequently eliminating the ability to describe adaptation in a generalized sense. Cowin and associates[64] performed cubic approximations of the theory of internal remodeling, and performed numerous studies attempting to establish possible values of the remodeling coefficients. Cowin and associates[6] also described a computational approach to the theory of surface remodeling enroute to predicting *in vivo* values for surface remodeling rate coefficients.

Employing the surface remodeling theory established by Cowin and Van Buskirk,[67] Cowin and Firoozbakhsh[68] presented a variety of theoretical predictions of surface remodeling in the diaphysis of long bones. For example, both endosteal and periosteal surfaces can move in either direction, in or out, in the same or opposing directions. It is possible for the medullary canal to fill completely, subsequently causing the endosteal surface to vanish. They proposed that the limitations of Cowin and Van Buskirk[67] were attributable to their assumption that the movement of the periosteal and endosteal surfaces was small.

The Cell Biology-Based Model

Hart et al.[11,69] utilized the adaptive elasticity model proposed by Cowin and co-workers to develop a computational model used to predict strain-history-dependent remodeling in long bones. Rather than follow the mechanical phenomenological approach of the adaptive elasticity theory, the model developed the remodeling rate constants in terms of biological parameters including the number of different cells present and their average daily activity. The basic premise of the model was that since bone is both resorbed and formed by cells that line the bony surfaces, bone remodeling is the manifestation of surface cellular processes. Hart's computational model was constructed around the techniques of the finite element method. The model was extended to incorporate the influence of material maturation (i.e., the material added to the surface of the bone was allowed to mature with time). Results were in agreement with the available analytical results and added to the importance of coupled remodeling effects not examined previously.

Strain Energy Density (SED) Theory of Adaptive Bone Remodeling

Huiskes and co-workers[37] proposed an alternative to the formulation of the theory of adaptive elasticity utilizing the strain energy density function as the remodeling signal rather than the strain tensor. As a scalar, the SED (U) represents the deformational energy available at any point:

$$U = \tfrac{1}{2}\, \varepsilon_{ij}\sigma_{ij} \tag{2.4}$$

where σ_{ij} and ε_{ij} represent the local stress and strain tensors, respectively. The driving mechanism for adaptive activity is assumed to be the aberration between the actual SED (U) and a site-specific homeostatic equilibrium SED, (U_n). Following a suggestion from Carter,[70] Huiskes assumed bone to be "lazy." In effect, he assumed that a certain threshold level, s, in overloading or underloading must be exceeded before bone reacts (Fig. 2.7). Mathematically, the internal remodeling rule becomes:

$$dE/dt = C_e\big(U-(1+s)U_n\big); \quad U > (1+s)U_n$$

$$= 0; \qquad\qquad (1-s)U_n \leq U \leq (1+s)U_n \tag{2.5}$$

$$= C_e\big(U-(1-s)U_n\big); \quad U < (1-s)U_n$$

where E is the elastic modulus of the element in question and C_e is the internal remodeling rate constant to be determined experimentally. External remodeling is represented by a similar modified formula, such that dx/dt exemplifies the rate of surface growth normal to the surface.

In the absence of a "lazy zone" (i.e., s = 0), internal remodeling is represented by:

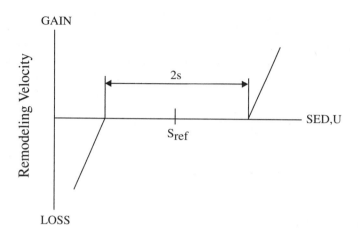

FIGURE 2.7 The adaptive remodeling rate as a function of strain energy density (SED) with threshold (s). (*Source*: Adapted from Huiskes, R. et al., *J. Biomech.*, 20, 1135, 1987. With permission from Elsevier Science.)

$$dE/dt = C_e(U - U_n) \qquad (2.6)$$

while the external relocation of surface nodes takes the form:

$$dx/dt = C_x(U - U_n) \qquad (2.7)$$

To date, the remodeling rates (i.e., C_e, and C_x) have not been established and are defined arbitrarily. Consequently, only the end result is deemed realistic.

Theory of Self-Optimization or Bone Maintenance

Fyhrie and Carter[10] advanced a theory suitable in principle to describe the self-optimization capabilities of bone Wolff proposed mathematically. They postulated that bone would adapt its apparent density and trabecular orientation locally for any loading environment in order to normalize a predestined effective stress value. The proposed bone remodeling objective was approximated using two independent measures of structural integrity: one based on strain energy; the other based on failure stress. The strain energy density (SED) principle optimized the stiffness while strength was optimized via the failure stress principle. Both measures were capable of predicting the orientation of trabeculae consistent with the trajectory hypotheses of Roux, von Meyer, Culmann, and Wolff and define stress measures that can be used to predict apparent density.[10] Each optimization approach predicted that at equilibrium the apparent density of cancellous bone is related to applied stress via:

$$\rho \propto \sigma_{eff}^c \qquad (2.8)$$

where σ_{eff} is an effective stress measure. The predicted value of the exponent C depends upon whether stiffness of strength optimization is assumed. Fyhrie and Carter[71] discovered that the apparent density was proportional to the square root of an effective stress ($C \approx 1/2$), when optimizing strength. By contrast, the strain energy density approach or strain energy density principle conveyed an apparent density proportional to the cube root of an effective stress squared ($C \approx 2/3$).

As introduced initially,[10] the SED principle did not optimize the strain energy density of the bone tissue but rather it optimized the apparent strain energy density of the continuum representation of the cancellous bone. The average "true" strain energy density of bone tissue, U_b, was subsequently related to the apparent SED, U, via:[72]

$$U_b = U/v \tag{2.9}$$

where v represents the volume fraction of mineralized bone tissue.[73] The argument for using U_b as the objective function was that "one would expect cellular reactions to stress to be a result of mineralized tissue strain energy only rather than an average of strain energy over mineralized tissue and marrow spaces."[72] Fyhrie and Carter[71] utilized this predictive theory to contrive the bone density distribution concordant with that found in the normal femoral head when a loading condition representing the single limb stance of gait was applied.

Whalen et al.[74] and Carter et al.[72] expanded the single-load approach for predicting bone density to encompass the multiple-loading history of bone over a predefined period. This approach characterized the bone loading histories for an "average day" in terms of stress magnitudes or cyclic strain density and the number of loading cycles. The assumption that bone mass is adjusted in response to strength or energy considerations enabled relationships between the local bone apparent density and loading history to be established.

Beaupré et al.[34] sought to extend the bone maintenance theory developed by Carter and colleagues into a time-dependent modeling/remodeling theory. The daily mechanical stimulus ψ_b was defined as:

$$\psi_b = (\rho_c/\rho)^2 \psi \tag{2.10}$$

where ρ_c is the density of cortical bone (assumed to be approximately equal to the density of mineralized tissue); ρ is the apparent density (mineralized tissue mass per total tissue volume); and ψ represents the daily stress stimulus measured at the continuum level. The tissue level remodeling error, e, in the following form, represents the difference between the actual tissue level stress stimulus and the tissue level attractor state stress stimulus:

$$e = \psi_b[\psi - (\rho/\rho_{AS})^2 \psi_{AS}]/\psi \tag{2.11}$$

where ρ_{AS} and ψ_{AS} denote the bone density and stress stimulus of the attractor state (AS), respectively. This error constitutes the driving force for bone remodeling.

Intertwining the time-dependent remodeling rules governing the theories of Huiskes, and Beaupré, and the self-optimization or bone maintenance theory,[71] Weinans et al.[38] sought to obtain a better understanding of the behavior of strain-adaptive bone remodeling in combination with FE models. In particular, the stability and convergence behavior of the remodeling rule were investigated in relation to the characteristics of the FE mesh. Hence the remodeling objective took on the following form:

$$\partial\rho/\partial t = B(U_a/\rho - k); \quad 0 < \rho < \rho_{cb} \tag{2.12}$$

where, $\rho = \rho(x, y, z)$ is the apparent density, B and k are constants, and

$$U_a = (1/n)\sum_{i=1}^{n} U_i \tag{2.13}$$

This process was considered to have *converged* when $\partial\rho/\partial t$ attained zero value according to Eq. (2.6) or when the density secured a minimal or maximal value. The stimulus, as a rule, was measured per element. Each element in principle possessed three possible paths of convergence en route to remodeling equilibrium: (1) the bone absorbed completely ($\rho = 0$); (2) the bone became cortical ($\rho = \rho_{cb}$); or (3) the bone remained cancellous with an apparent density satisfying $E = c\rho^\gamma$. Based on their results, the following hypotheses were drawn: (1) that bone is indeed a self-optimizing material that produces a self-similar trabecular morphology, a fractal, in a chaotic process of self-organization, whereby uniform SED per unit mass or a similar mechanical signal is an attractor; (2) that the morphology has qualities of minimal weight; and (3) that its morphological and dimensional characteristics depend on the local loading characteristics, the maximal degree of mineralization, the sensor density, and the attractor value.

It should be noted that these models are empirical, not physiological. They may be used to estimate the outcome of a remodeling process, but provide little or no explanation about the operation of remodeling process. Nonetheless, the aforementioned models provide an evolution of adaptive remodeling techniques.

Unstable Behavior

The literature is scarce regarding the stability of the proposed theoretical bone remodeling simulations. The lack of an analytical solution for stresses, in a medium where material properties vary with position, acts as a limitation to the study of these models. Harrigan and Hamilton[75] sought the origin of unstable bone remodeling simulations mathematically, using strain-energy-based remodeling rules in an attempt to assess whether the unstable behavior was due to the mathematical rules proposed to characterize the process or to the numerical approximations used to exercise the mathematical predictions. The physiologic interpretation indicated that the instabilities that occur in some remodeling simulations are due, at least in part, to the mathematical characterization of bone remodeling. In addition, the behavior of the observed instabilities is not present *in vivo*. Consequently, the cause of this unstable behavior is most likely not attributed to natural remodeling processes.

Carter et al.[35] observed that when their simulation of femoral bone remodeling was allowed to progress past the first few iterations, "The method employed appeared to converge toward a condition in which most elements will either be saturated … or be completely resorbed." Recently, Weinans et al.[38,76] demonstrated, confirmed by others,[75,77] that previous bone remodeling implementations tend toward discontinuous density patterns (Fig. 2.8). In the vicinity of the applied loads, elements predict alternating patterns of high and low density, resembling the pattern of a checkerboard. Jacobs et al.,[77] in an attempt to eliminate the spurious near-field discontinuities, while maintaining anatomically correct far-field discontinuities, implemented a "node-based" technique.

Fyhrie and Schaffler,[78] in the same vein, sought to improve spatial stability via a revised phenomenological theory of bone remodeling. They cite that bone remodeling theories are often based on the common assumption that the changes in bone structure in response to an *error* signal are *adaptive*, and therefore bring about a reduction in error. They criticized that under these assumptions, the basic formulation of the remodeling problem is to adapt the structure to make the error approach zero. Consequently, this formulation will not converge to an optimal bone structure unless the error function is specifically designed to do so. If, however, the optimality is defined as zero signal error at each point in the bone, this formulation does result in an optimal solution. En route to the development of a new remodeling theory, the following distinctions were made. The apparent density was identified as the controlling variable, while the controlled variable was a function of the apparent strain, denoted $M(E)$. The controlling and controlled variables were defined as those which the bone cells can directly modify, and those which measure the ability of bone to adapt to the current need, respectively. Although the precise form of the function $M(E)$ is not known presently, it is considered the homeostatic value of apparent density attained by bone subjected to constant strain. The fact that the function is not necessarily zero as the strain magnitude goes to zero accounts for the biological factors which prevent the total disappearance of bone tissue. The fundamental character of the remodeling equation was exponential, consistent with experimental observations of changes during disuse, after hip replacement surgery, and during growth and aging. Fyhrie and Schaffler were able to demonstrate that the model is stable temporally, and more spatially stable than some models published previously.

Causal Mechanisms

The origin and function of adaptive remodeling have been debated extensively. The feedback mechanism by which bone tissue senses the change in load environment and initiates the deposition or resorption of bone is not understood.[15] Undoubtedly, mechanical factors play an important role in remodeling;

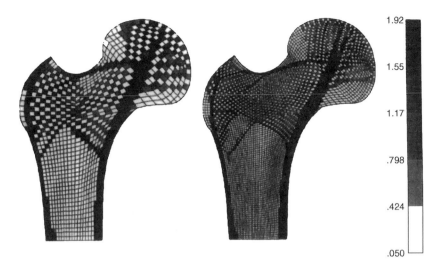

FIGURE 2.8   Predicted checkboard density distribution characteristic of the traditional element-based bone remodeling algorithms. *Source*: Jacobs, C.R. et al., *J. Biomechanics*, 28, 449, 1995. With permission.

inactivity results in widened Haversian canals and porotic bone, while stresses result in a more solid compactum. Recent investigations have explored the biological response of bone to mechanical loading at the cellular level, but the precise mechanosensory system that signals bone cells to deposit or resorb tissue has not been identified.[15]

Numerous recorded observations suggest that bone cells *in situ* are capable of responding to mechanical stimuli and do so in a predictable fashion (i.e., Wolff's law). Experimental limitations often hinder such investigations at the cellular level. A major constraint of *in vitro* organ culture conditions is that the cultured structures are complex and composed of heterogeneous cell populations.[15] Although *in vivo* loading conditions may be approximated, extracting satisfactory information from these models regarding individual cell behaviors is laborious. Nonetheless, experimental procedures have implicated different mechanisms for adaptive bone remodeling.

Debate is ongoing as to the mechanical signal to which bone cells respond. The question with regard to the causation of adaptation — stress or strain? — is intrinsically difficult to answer due to their direct proportionality. Although closely related, their relationship in a nonhomogeneous and anisotropic material such as bone is altogether variable. Stress is an abstract concept, the components of which must be deduced from measurements of load, or from measurements of strain and elastic constants.[22] In addition, stress is defined based solely on its effects. Strain may be measured directly via strain gauges, or calculated from measured displacement fields as the symmetric part of the displacement gradient. Cowin[79] states that the reason bones sense strain rather than stress is that strain is a primary, directly measurable physical quantity, whereas stress is not.

The advent of *in vivo* strain gauging techniques that permit direct measurement of bone deformation prompted a series of experiments to define and quantify the nature of the relationship between mechanical loading and bone remodeling. The results of experiments employing *in vivo* strain gauge techniques spanning a 15-year period have been used to support the contention that bone senses and responds to strain rather than stress.[22] Numerous strain-gauged animal experiments have been performed by Lanyon and Rubin[23,80-82] to assess the relationship between bone tissue response and tissue level strain magnitude. Experimentation has confirmed that bone remodeling is responsive to dynamic strains within the matrix, manifesting a progressively increasing osteogenic response to progressively increased loading.[83] These experiments consisted of depriving a bone of its normal loading *in vivo*, while interrupting the subsequent disuse by daily intermittent loading. Strains of less than 0.001 were associated with the loss of bone, whereas elevated strains resulted in a proportional increase in bone area.

The components of a dynamic strain regime which influence remodeling behavior have yet to be characterized completely. However, they appear to include peak strain magnitude, strain rate, and strain distribution.[82] In combination, these factors play some role in producing an effective strain stimulus. The reader is referred to Burr[84] and Martin and Burr[22] for a complete description of the aforementioned potential mechanical stimuli. Although little hard experimental evidence suggests that strain energy provides the adaptive signal, it is often used theoretically to model the development and adaptation of bone and cartilage.[37,55,74,85] Strain energy is proportional to the product of stress and strain. However, strain energy possesses two characteristics distinguishing it from both stress and strain: (1) it is a scalar rather than a tensor; and (2) it is always positive regardless of whether the loads are tensile or compressive.[22] Consequently, strain energy represents a less complex variable as compared to strain, but more information is needed.[22]

Although the majority of theories emphasize the mechanical aspects of bone adaptation, the process cannot be reduced to a purely mechanical form. It is highly possible that cells are not sensitive to stress or strain, but to another factor (i.e., electrical potential) generated via the stress or strain fields. A number of chemical reactions supplement the bone remodeling process. Although the mechanism responsible for these reactions continues to elude researchers, there are two promising candidates: one electrical, the other chemical.

At the cellular level, stretch-activated ion channels transduce mechanical strain into an ion flux or an electrical response.[86] This mechanism is activated when the cell membrane is strained, thus developing a preferential passageway for the transit of specific ions. The aforementioned cellular-level strains are classified as highly localized at the cell lacunae level; by contrast, tissue level strains represent macroscopic strain averages over a significant volume of bone tissue. In 1953, the work of Fukada and Yasuda[87] led to the hypothesis that strain-related electrical potentials mediate the adaptive response. The aforementioned theory of piezoelectricity in cortical bone led Gjelsvik[88,89] to derive mathematically a theory of mechanically adaptive surface remodeling. This theory proposed that resorption would occur systemically on all bone surfaces, while apposition in proportion to the surface charge counterbalanced this tendency.

Utilizing the constants derived by Fukada and Yasuda,[87] Gjelsvik observed the effects of alterations in mechanical usage, and the classical problem of the flexural neutralization in an angulated bone.[89] As interpreted by Martin,[22] the resulting data implied that the collagen molecules possess a left-hand twist on one side of the body and a right-hand twist on the other. This, however, is not feasible since all naturally occurring collagen has the same direction of twist.[90,91] Consequently, the likelihood of piezoelectricity governing the adaptive response in bone is probably not as great as initially anticipated.

Subsequent investigations suggest that the physiologically significant strain generated potential (SGP) in bone is not piezoelectricity, but electrical potential of electrokinetic origin.[92,93] When pressure differentials between two sites in bone tissue elicit flow of the charged fluid in bone channels, a streaming current which gives rise to SGP is established. The potential difference or streaming potential between the two sites may, in turn, be measured. Hence, transient pressures and fluid flow have been cited as potential candidates governing adaptive bone remodeling. Jendrucko et al.[94] evaluated the relationship between applied compressive stress and the pressure exerted on an osteocyte. Axial compressive loading of an osteon was shown to induce radial flow.[95,96]

2.4 Soft Tissue Remodeling

Brickley-Parsons et al.[97] suggest that while Wolff's law was formulated originally to describe the adaptive response of bone to externally applied mechanical forces, there is no *a priori* reason why the same biological principles do not apply to other skeletal structures whose major functions are also mechanical in nature. Perhaps the most well-known example is the hypertrophy of muscle following athletic training. In contrast to the extensive work on bones, very little has been done on modeling the relationship of stress, strain, and growth in soft tissues. This has been attributed to the fact that soft tissues typically exhibit large elastic deformations under physiological loading.[98] In recent years, however, phenomena suggesting response to the change in mechanical conditions have been observed in soft biological tissues.

Aortic walls in hypertensive rats, for example, increase their thickness as if the hypertrophy maintained the circumferential stress at a similar level to that in normotensive animals.[99-101] At the cellular level, myocardial cells have been shown to atrophy in response to a month-long reduction in cardiac work as the result of an artificial assist device.[102] Matsumoto et al.[103] suggest that dimensional changes such as wall thickening observed in hypertensive aortas and overloaded left ventricles seem to occur to maintain the mechanical stresses developed under *in vivo* conditions, approximating the same level as that in normal tissues and organs.

Inspired by the fact that growth and remodeling in tissues may be modulated by mechanical factors such as stress, Rodriguez et al.[98] proposed a general continuum formulation for finite volumetric growth in soft elastic tissues. The shape change of an unloaded tissue during growth was described by a mapping, analogous to the deformation gradient tensor. This mapping was decomposed into a transformation of the local zero-stress reference state and an accompanying elastic deformation that ensured the compatibility of the total growth deformation. Residual stresses arose from the elastic deformation. With a thick-walled hollow cylinder of incompressible, isotropic hyperelastic material as an example, the mechanics of left ventricular hypertrophy were analyzed. Results indicate that transmurally uniform pure circumferential growth, which may be similar to eccentric ventricular hypertrophy, changes the state of residual stress in the heart wall.

Yamamoto et al.[104] investigated the effects of stress shielding on the mechanical properties of the rabbit patellar tendon. Stress shielding was accomplished by stretching a stainless steel wire installed between the patella and tibial tubercle, thus releasing the tension in the patellar tendon completely. Significant alterations in the mechanical properties of the patellar tendon were observed as the result of stress shielding. It decreased the tangent modulus and tensile strength to 9% of the control values after 3 weeks. There was a 131% increase in the cross-sectional area and a 15% decrease in the tendinous length. Histological studies revealed that the stress shielding increased the number of fibroblasts while decreasing the longitudinally aligned collagen bundles.

2.5 Summary

The skeleton's capacity to withstand external loading is achieved and maintained because the adaptive remodeling of bone tissue is both sensitive and responsive to the functional demands placed upon it. Numerous attempts to quantify the adaptive phenomena of bone have been reported in the literature. Qualitative predictions require that the internal mechanical load on the bone structure be determined accurately in terms of stresses and strains, for which the finite element method (FEM) has proven an effective tool.[38,105] Mathematical bone remodeling theories, in conjunction with finite element models, enable bone formation and resorption patterns in realistic bone structures to be predicted quantitatively.

To date, the precise mechanism underlying the functional adaptation of bone tissue continues to elude researchers. It appears, however, to involve simultaneous cell-controlled mechanical, bioelectric, and biochemical processes.[15] Numerous candidate mechanosensory transduction mechanisms, ranging from mechanical to electrical in nature, have been proposed.

The capability of remodeling algorithms to yield more realistic density distributions and external configurations continues to improve. Undoubtedly, an accurate representation of the bone remodeling process would provide significant clinical benefits. The possibilities for artificial joint replacement, pre- and post-clinical testing, and clinical research generally are endless.

Acknowledgment

This work was supported in part by the Whitaker Foundation and the University of Iowa Spine Research Fund.

References

1. Galilei, G., *Discorsi e Dimostrazioni Matematiche Intorna a Due Nuove Scienze*, Macmillan, New York, 1914, 158.
2. Wolff, J., *Das Gezetz der Transformation der Knochen*, Hirschwald, Berlin, 1892.
3. Frost, H.M., *Bone Modeling and Skeletal Remodeling Errors*, Charles C Thomas, Springfield, IL, 1973.
4. Frost, H.M., *Bone Remodeling and Its Relation to Metabolic Disease*, Charles C Thomas, Springfield, IL, 1973.
5. Adams, D., Ph.D. thesis, University of Iowa, Iowa City, 1996.
6. Cowin, S.C., Hart, R.T., Balser, J.R., and Kohn, D.H., *J. Biomechanics*, 18, 665, 1985.
7. Frost, H.M., *The Laws of Bone Structure*, Charles C Thomas, Springfield, IL, 1964
8. Frost, H.M., *Intermediary Organization of the Skeleton*, CRC Press, Boca Raton, FL, 1986.
9. Frost, H.M., *Bone Miner.*, 2, 73, 1987.
10. Fyhrie, D.P. and Carter, D.R., *J. Orthoped. Res.*, 4, 304, 1986.
11. Hart, R.T., Davy, D.T., and Heiple, K.G., *J. Biomechanical Eng.*, 106, 342, 1984.
12. Hart, R.T., Davy, D.T., and Heiple, K.G., *Calc. Tissue Int.*, 36, S104, 1984.
13. Roesler, H., *J. Biomechanics*, 20, 1025, 1987.
14. Rubin, C.T., in *Non-Cemented Total Hip Arthroplasty*, Fitzgerald, R., Ed., Raven Press, New York, 1988.
15. Cowin, S.C., Moss-Salentijn, L., and Moss, M.L., *J. Biomechanical Eng.*, 113, 191, 1991.
16. Martini, F.H., *Fundamentals of Anatomy and Physiology*, Prentice-Hall, Englewood Cliffs, NJ, 1995.
17. Weiss, L., in *Cell and Tissue Biology*, Elsevier, New York, 1983.
18. Bouvier, M., in *Bone Mechanics*, Cowin, S.C., Ed., CRC Press, Boca Raton, FL, 1989.
19. Parfitt, A.M., in *Bone Histomorphometry: Techniques and Interpretations*, CRC Press, Boca Raton, FL, 1983.
20. Hayes, W.C., in *Basic Orthopaedic Biomechanics*, Mow, V. and Hayes, W., Eds., Raven Press, New York, 1991.
21. Sevitt, S., *Bone Repair and Fracture Healing in Man*, Churchill Livingstone, New York, 1981, 15.
22. Martin, R.B. and Burr, D.B., *Structure, Function, and Adaptation of Compact Bone*, Raven Press, New York, 1989.
23. Lanyon, L.E., *Calc. Tissue Int.*, 36, S56, 1984.
24. Huiskes, R., in *Basic Orthopaedic Biomechanics*, Mow, V. and Hayes, W., Eds., Raven Press, New York, 1991.
25. Hedley, A.K. et al., *Clin. Orthoped.*, 163, 300, 1982.
26. Cook, S.D., Thomas, K.A., and Haddad, R.J., *Clin. Orthoped.*, 234, 90, 1986.
27. Goldstein, S.A., in *Bone Biodynamics in Orthodontic and Orthopedic Treatment*, Carlson, D. and Goldstein, S., Eds., Center for Human Growth and Development, Ann Arbor, MI, 1992.
28. Meade, J.B., in *Bone Mechanics*, Cowin, S.C., Ed., CRC Press, Boca Raton, FL, 1989.
29. Chamley, J., *Clin Orthoped.*, 95, 9, 1973.
30. Almby, B. and Hierton, T., *Acta Orthopaed. Scand.*, 53, 397, 1982.
31. Cotterill, G.A., Hunter, G.A., and Tile, M., *Clin. Orthoped.*, 163, 120, 1982.
32. Hamilton, L.R., *J. Bone Jt. Surg.*, 64A, 740, 1982.
33. Lewallen, D.B., Bryan, R.S., and Peterson, L.F.A., *J. Bone Jt. Surg.*, 66A, 1211, 1984.
34. Beaupré, G.S., Orr, T.E., and Carter, D.R., *J. Orthoped. Res.*, 8, 662, 1990.
35. Carter, D.R., Orr, T.E., and Fyhrie, D.P., *J. Biomechanics*, 22, 231, 1989.
36. Cowin, S.C., *J. Biomechanics*, 20, 1111, 1987.
37. Huiskes, R. et al., *J. Biomechanics*, 20, 1135, 1987.
38. Weinans, H., Huiskes, R., and Grootenboer, H.J., *J. Biomechanics*, 25, 1425, 1992.
39. Brown, T.D., Pederson, Gray, M.L., Grant, R.A., and Rubin, C.T., *J. Biomechanics*, 23, 893, 1990.
40. Beaupré, G.S., Orr, T.E., and Carter, D.R., *J. Orthoped. Res.*, 8, 651, 1990.

41. Orr, T., Beaupré, G., Fyhrie, D., Schurman, D., and Carter, D., Application of bone remodeling theory to femoral and tibial prosthetic components, in *Proceedings, 34th Annual Meeting, Orthopaedic Research Society*, 1988.
42. Orr, T., Beaupré, G., Carter, D., and Schurman, D., *J. Arthroplasty*, 5, 191, 1990.
43. Huiskes, R., *Acta Orthopaed. Scand.*, 51, 185S, 1980.
44. Miller, J.E. and Kelebay, L.C., *Orthoped. Trans.* 5, 380, 1981.
45. Chen, P.S. et al., *Clin. Orthoped. Rel. Res.*, 176, 24, 1983.
46. Dallante, P. et al., in *Biological and Biomechanical Performance of Biomaterials*, Christel, P. et al., Eds., Elsevier, Amsterdam, 1986.
47. Turner, T.M., Sumner, D.R., Urban, D.P., and Rivero, G.J.O., *J. Bone Jt. Surg.*, 68A, 1396, 1986.
48. Huiskes, R., Heck, J.V., Walker, P.S., Greene, D.J., and Nunamaker, D., *Finite Elements in Biomechanics*, University of Arizona, Tucson, 1980.
49. Huiskes, R. and Nunamaker, D., *Calc. Tissue Int.*, 36, S110, 1984.
50. Levenston, M.E., Beaupré, G.S., Schurman, D.J., and Carter, D.R., *J. Arthroplasty*, 8, 595, 1993.
51. Rapperport, D.J., Carter, D.R., and Schurman, D.J., *J. Orthoped. Res.*, 3, 435, 1985.
52. Goel, V.K. and Seenivasan, G., *IEEE Eng. Med. Biol.*, 13, 508, 1994.
53. Seenivasan, G., M.S. thesis, University of Iowa, Iowa City, 1993.
54. Goel, V.K., Kong, W., Han, J.S., Weinstein, J.N., and Gilbertson, L.G., *Spine*, 18, 1531, 1993.
55. Ramirez, S.A., M.S. thesis, University of Iowa, Iowa City, 1994.
56. Hegedus, D.H. and Cowin, S.C., *J. Elasticity*, 6, 337, 1976.
57. Grosland, N.M., Goel, V.K., Grobler, L.J., and Griffith, S.L., Adaptive internal bone remodeling of the vertebral body following an anterior interbody fusion: a computer simulation, *ISSLS* (Singapore), 1997.
58. Weinans, H. et al., *J. Bone Jt. Surg.*, 11, 500, 1993.
59. Koch, J.C., *Am. J. Anat.*, 21, 177, 1917.
60. Andre, N., *L'orthopedie ou L'art de Prevenir et de Corriger dans les Enfans*, Paris, 1741.
61. D'Arcy-Thompson, W., *On Growth and Form*, Cambridge University Press, Cambridge, 1942.
62. Pauwels, F., *Biomechanics of the Locomotor Apparatus: Contributions on the Functional Anatomy of the Locomotor Apparatus*, Springer-Verlag, Berlin, 1980.
63. Kummer, B.K. *Biomechanics of Bone*, Prentice-Hall, Englewood Cliffs, 1972.
64. Firoozbakhsh, K. and Cowin, S.C., *J. Biomechanical Eng.*, 103, 246, 1981.
65. Cowin, S.C. and Hegedus, D.H., *J. Elasticity*, 6, 313, 1976.
66. Cowin, S.C. and Nachlinger, R.R., *J. Elasticity*, 8, 285, 1978.
67. Cowin, S.C. and Van Buskirk, W.C., *J. Biomechanics*, 11, 313, 1978.
68. Cowin, S.C. and Firoozbakhsh, K., *J. Biomechanics*, 14, 471, 1981.
69. Hart, R.T., Davy, D.T., and Heiple, K.G., *J. Biomechanical Eng.*, 104, 123, 1982.
70. Carter, D.R., *Calc. Tissue Int.*, 36, S19, 1984.
71. Fyhrie, D.P. and Carter, D.R., *J. Biomechanics*, 23, 1, 1990.
72. Carter, R.R., Fyhrie, D.P., and Whalen, R.T., *J. Biomechanics*, 20, 785, 1987.
73. Carter, R.R., Fyhrie, D.P., and Whalen, R.T., Mathematical models for predicting bone density from stress history, 10th Annual Conference, American Society for Biomechanics, Montreal, 1986.
74. Whalen, R.T., Carter, D.R., and Steele, C.R., *J. Biomechanics*, 21, 825, 1988.
75. Harrigan, T.P. and Hamilton, J.J., *J. Biomechanics*, 25, 477, 1992.
76. Weinans, H., Huiskes, R., and Grootenboer, H.J., *Trans. 26th Orthoped. Res. Soc.*, 15, 78, 1990.
77. Jacobs, C.R., Levenston, M.E., Beaupré, G.S., Simon, J.C., and Carter, D.R., *J. Biomechanics*, 28, 449, 1995.
78. Fyhrie, D.P. and Schaffler, M.B., *J. Biomechanics*, 28, 135, 1995.
79. Cowin, S.C., *Calc. Tissue. Int.*, 36, S98, 1984.
80. Rubin, R.C. and Lanyon, L.E., *J. Bone Jt. Surg.*, 66A, 397, 1984.
81. Rubin, R.C. and Lanyon, L.E., *J. Orthoped. Res.*, 5, 300, 1987.
82. Lanyon, L.E., *J. Biomechanics*, 20, 1083, 1987.

83. Rubin, C.T. and Lanyon, L.E., *Calc. Tissue Int.*, 37, 411, 1985.

84. Burr, D.B., in *Bone Biodynamics in Orthodontic and Orthopedic Treatment*, Carlson, D. and Goldstein, S., Eds., Center for Human Growth and Development, Ann Arbor, MI, 1992.

85. Wong, M. and Carter, D.R., *Bone*, 11, 127, 1990.

86. Sachs, F., *Crit. Rev. Biomed. Eng.*, 16, 141, 1988.

87. Fukada, E. and Yasuda, I., *J. Phys. Soc. Japan*, 12, 1158, 1957.

88. Gjelsvik, A., *J. Biomechanics*, 6, 69, 1973.

89. Gjelsvik, A., *J. Biomechanics*, 6, 187, 1973.

90. Martin, R.B., *in Electrical Properties of Bone and Cartilage*, Brighton, C. et al., Eds., Grune and Stratton, New York, 1979.

91. Martin, R.B., *J. Biomechanics*, 12, 55, 1979.

92. Gross, D. and Williams, W.S., *J. Biomechanics*, 15, 277, 1982.

93. Pollack, S.R., Salzstein, R., and Pienkowski, D., *Ferroelectronics*, 60, 297, 1984.

94. Jendrucko, R.J., Hyman, W.A., Newell, P.H., Jr., and Chakraborty, B.K., *J. Biomechanics*, 9, 87, 1976.

95. Piekarski, K. and Munro, M., *Nature*, 269, 80, 1977.

96. Munro, M. and Piekarski, K., *J. Appl. Mech.*, 99, 218, 1977.

97. Brickley-Parsons, D. and Glimcher, M.J., *Spine*, 9, 148, 1984.

98. Rodriguez, E.K., Hoger, A., and McCulloch, A.D., *J. Biomechanics*, 27, 455, 1994.

99. Vaishnav, R.N. et al., *J. Biomechanical Eng.*, 112, 70, 1990.

100. Wolinsky, H., *Circ. Res.*, 28, 622, 1971.

101. Wolinsky, H., *Circ. Res.*, 30, 301, 1972.

102. Nakamura, T. et al., *Biomed. Mater. Eng.*, 2, 139, 1992.

103. Matsumoto, T. and Kayashi, K., *J. Biomechanical Eng.*, 116, 278, 1994.

104. Yamamoto, N. et al., *J. Biomechanical Eng.*, 115, 23, 1993.

105. Huiskes. R. and Chao, E.Y.S., *J. Biomechanics*, 16, 385, 1983.

3

Techniques in the Dynamic Modeling of Human Joints with a Special Application to the Human Knee

Ali E. Engin
University of South Alabama

3.1   Introduction

Joints in the human skeletal structure can be roughly classified into three categories according to the amount of movement available at the joint. These categories are named synarthroses (immovable), amphiarthroses (slightly movable) and diarthroses (freely movable). The skull sutures represent examples of synarthrodial joints. Examples of amphiarthrodial joints are junctions between the vertebral bodies and the distal tibiofibular joint. The main interest of this chapter is the biomechanic modeling of the major articulating joints of the upper or lower extremities that belong to the last category, the diarthroses. In general, a diarthrodial joint has a joint cavity which is bounded by articular cartilage of the bone ends and the joint capsule.

The bearing surface of the articular cartilage is almost free of collagen fibers and is thus true hyaline cartilage. From a biomechanics view, articular cartilage may be described as a poroelastic material composed of solid and fluid constituents. When the cartilage is compressed, liquid is squeezed out, and, when the load is removed, the cartilage returns gradually to its original state by absorbing liquid in the process. The time-dependent behavior of cartilage suggests that articular cartilage might also be modeled as a viscoelastic material, in particular, as a Kelvin solid.

The joint capsule encircles the joint and its shape is dependent on the joint geometry. The capsule wall is externally covered by the ligamentous or fibrous structure (fibrous capsule) and internally by synovial membrane which also covers intra-articular ligaments. Synovial membrane secretes the synovial fluid which is believed to perform two major functions. It serves as a lubricant between cartilage surfaces and also carries out metabolic functions by providing nutrients to the articular cartilage. The synovial fluid is non-Newtonian in behavior, i.e., its viscosity depends on the velocity gradient. Cartilage and synovial fluid interact to provide remarkable bearing qualities for the articulating joints. More information on properties of articular cartilage and synovial fluid can be found in a book chapter written by the author in 1978.[8]

Structural integrity of the articulating joints is maintained by capsular ligaments and both extra- and intra-articular ligaments. Capsular ligaments are formed by thickening of the capsule walls where functional demands are greatest. As the names imply, extra- and intra-articular ligaments at the joints reside external to and internal to the joint capsule, respectively. Extra-articular ligaments have several shapes, e.g., cord-like or flat, depending on their locations and functions. These types of ligaments appear abundantly at the articulating joints. However, only the shoulder, hip, and knee joints contain intra-articular ligaments. The cruciate ligaments at the knee joint are probably the best known intra-articular ligaments. Further information about the structure and mechanics of the human joints is available in Reference 1.

Returning to the modeling aspects of the articulating joints, in particular kinematic behavior of the joints, we can state that in each articulating human joint, a total of six degrees-of-freedom exist to some extent. One must emphasize the point that degrees-of-freedom used here should be understood in the sense the phrase is defined in mechanics, because the majority of the anatomists and the medical people have a different understanding of this concept; e.g., both Steindler[34] and MacConaill[26] imply that the maximum number of degrees-of-freedom required for anatomical motion is three.

Major articulating joints of the human have been studied and modeled by means of joint models possessing single and multiple degrees-of-freedom. Among the various joint models the hinge or revolute joint is probably the most widely used articulating joint model because of its simplicity and its single degree-of-freedom character. When the articulation between two body segments is assumed to be a hinge type, the motion between these two segments is characterized by only one independent coordinate which describes the amount of rotation about a single axis fixed in one of the segments. Although the most frequent application of the hinge joint model has been the knee, the other major joints have been treated as hinge joints in the literature, sometimes with the assumption that the motion takes place only in a particular plane, especially when the shoulder and the hip joints are considered.

When the degrees-of-freedom allowed in a joint model are increased from one to two, one obtains a special case of the three-degrees-of-freedom spherical or ball and socket joint. Two versions of this spherical joint which have received some attention in the literature. In the first version, no axial rotation of the body segment is allowed and the motion is determined by the two independent spherical coordinates ϕ and θ as shown in Fig. 3.1. In the second version, the axial rotation is allowed but the motion is restricted to a particular plane passing through the center of the sphere. Again, most of the major joints have been modeled by the two-degree-freedom spherical joint models by various investigators.

If we increase the degrees-of-freedom to three, we get the two obvious joint models, namely, the ball and socket joint model and the planar joint model. For the ball and socket joint model in addition to ϕ and θ, a third independent coordinate, ψ, which represents the axial rotation of one of the body segments, is introduced. The planar joint model, as the title suggests, permits the motion on a single plane and is characterized by two Cartesian coordinates of the instantaneous center of rotation and one coordinate, θ, defining the amount of rotation about an axis perpendicular to the plane of motion. Dempster[6] appears to be the first to apply the instant centers technique to the planar motion study of the knee joint.

The six-degrees-of-freedom joint (general joint) allows all possible motions between two body segments. A good example of a general joint is the shoulder complex, which exhibits four independent articulations among the humerus, scapula, clavicle, and thorax. Of course, at the shoulder complex, the six degrees-of-freedom refer to the motion of the humerus relative to the torso. If one considers the total

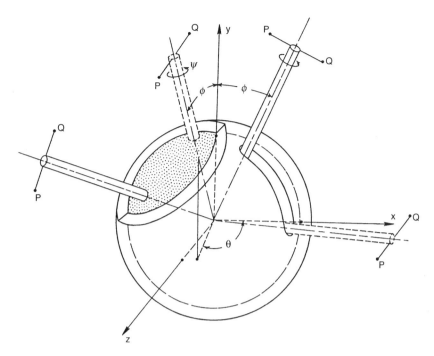

FIGURE 3.1   Spherical or ball and socket joint is illustrated. Figure displays both versions of the two degrees-of-freedom as well as the most general three-degrees-of-freedom spherical joint.

number of degrees-of-freedom for the motions executed by the various bones of the shoulder complex, one can easily reach a number higher than six, even with the proper consideration of various constraints present in the joint complex.

A substantial difficulty in theoretical modeling of human joints arises from the fact that the number of unknowns are usually far greater than the number of available equilibrium or dynamic equations. Thus, the problem is an indeterminate one. To deal with this indeterminate situation, optimization techniques have been employed in the past.[32,33] However, the selection of objective functions appears to be arbitrary, and justification for such minimization criteria is indeed debatable.

Another technique dealing with the indeterminate nature of joint modeling considers the anatomical and physiological constraint conditions together with the equilibrium or dynamic equations. These constraint conditions include the fact that soft tissues only transmit tensile loads while the articulating surfaces can only be subjected to compression. Electromyographic data from the muscles crossing the joint also provide additional information for the joint modeling effort. The different techniques used by various researchers mainly vary as to the method of applying these conditions. At one extreme, all unknowns are included in the equilibrium or dynamic equations. A number of unknown forces are then assumed to be zero to make the system determinate so that the reduced set of equations can be solved. This process is repeated for all possible combinations of the unknowns, and the values of the joint forces are obtained after discarding the inadmissible solutions.[5] At the other extreme, first the primary functions of all structures are identified and equations are simplified before they are solved.[30] A combination of these two techniques was also used to solve several quasi-static joint modeling problems.[3,4]

All models described in the previous paragraph are quasi-static. That is, the equilibrium equations together with the inertia terms are solved for a known kinematic configuration of the joint. Complexity of joint modeling becomes paramount when one considers a true dynamic analysis of an articulating joint structure possessing realistic articulating surface geometry and nonlinear soft tissue behavior. Because of this extreme complexity, the multisegmented models of the human body, thus far, have employed simple geometric shapes for their joints.

Although the literature[40] cites mathematical joint models that consider both the geometry of the joint surfaces and behavior of the joint ligaments, these models are quasi-static in nature, and employ the so-called inverse method in which the ligament forces caused by a specified set of translations and rotations along the specified directions are determined by comparing the geometries of the initial and displaced configurations of the joint. Furthermore, for the inverse method utilized in Reference 40, it is necessary to specify the external force required for the *preferred* equilibrium configuration. Such an approach is applicable only in a quasi-static analysis. For a dynamic analysis, the equilibrium configuration preferred by the joint is the unknown and the mathematical analysis is required to provide that dynamic equilibrium configuration.

In this chapter, a formulation of a three-dimensional mathematical dynamic model of a general two-body-segmented articulating joint is presented first. The two-dimensional version of this formulation subsequently is applied to the human knee joint to investigate the relative dynamic motion between the femur and tibia as well as the ligament and contact forces developed in the joint. This mathematical joint model takes into account the geometry of the articulating surfaces and the appropriate constitutive behavior of the joint ligaments. Representative results are provided from solutions of second-order nonlinear differential equations by means of the Newmark method of differential approximations and application of the Newton-Raphson iteration process. Next, to deal with shortcomings of the iterative method, alternative methods of solution of the same dynamic equations of the joint model are presented. With improved solution methods, the dynamic knee model is utilized to study the response of the knee to impact loads applied at any location on the lower leg. The chapter also deals with the question of whether the classical impact theory can be directly applied to dynamic joint models and its limitations. In addition, the two-body segmented joint model is extended to a three-body segmented formulation, and an anatomically based dynamic model of the knee joint which includes patello-femoral articulation is presented to assess patello-femoral contact forces during kicking activity.

3.2   General Three-Dimensional Dynamic Joint Model

The articulating joint is modeled by two rigid body segments connected by nonlinear springs simulating the ligaments. It is assumed that one body segment is rigidly fixed while the second body segment is undergoing a general three-dimensional dynamic motion relative to the fixed one. The coefficients of friction between the articulating surfaces are assumed to be negligible. This is a valid assumption due to the presence of synovial fluid between the articulating surfaces.[31] Accordingly, the friction force between the articulating surfaces will be neglected.

The main thrust of this section is the presentation of a mathematical modeling of an articulating joint defined by contact surfaces of two body segments which execute a relative dynamic motion within the constraints of ligament forces. Mathematical equations for the joint model are in the form of second-order nonlinear differential equations coupled with nonlinear algebraic constraint conditions. Solution of these differential equations by application of the Newmark method of differential approximation and subsequent usage of the Newton-Raphson iteration scheme will be discussed. The two-dimensional version of the dynamic joint model will be applied to the human knee joint under several dynamic loading conditions on the tibia. Results for the ligament and contact forces, contact point locations between the femur and tibia, and the corresponding dynamic orientation of the tibia with respect to femur will be presented.

Representation of the Relative Positions

The position of the moving body segment 1 relative to fixed body segment 2 is described by two coordinate systems as shown in Fig. 3.2. The inertial coordinate system (x, y, z) with unit vectors $\hat{i}$, $\hat{j}$ and $\hat{k}$ is connected to the fixed body segment and the coordinate system (x', y', z') with unit vectors $\hat{i}'$, $\hat{j}'$ and $\hat{k}'$ is attached to the center of mass of the moving body segment.

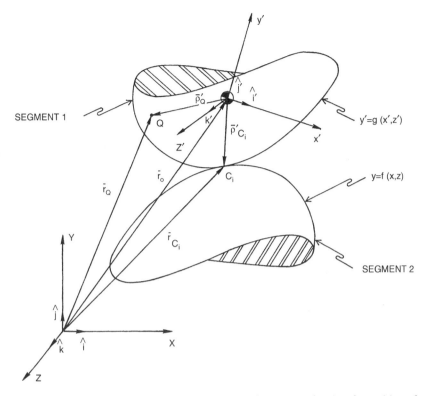

FIGURE 3.2 A two-body segmented joint is illustrated in three dimensions, showing the position of a point, Q, attached to the moving coordinate system (x', y', z').

The (x', y', z') coordinate system is also taken to be the principal axis system of the moving body segment. The motion of the (x', y', z') system relative to the fixed (x, y, z) system may be characterized by six quantities: the translational movement of the origin of the (x', y', z') system in the x, y, and z directions, and θ, ϕ, and ψ rotations with respect to the x, y, and z axes.

Let the position vector of the origin of the (x', y', z') system in the fixed system be given by (Fig. 3.2):

$$\bar{r}_o = x_o \hat{i} + y_o \hat{j} + z_o \hat{k} \tag{3.1}$$

Let the vector, $\bar{\rho}'_Q$ be the position vector of an arbitrary point, Q, on the moving body segment in the base $(\hat{i}', \hat{j}', \hat{k}')$. Let $\bar{r}_Q$ be the position vector of the same point in the base $(\hat{i}, \hat{j}, \hat{k})$. That is,

$$\bar{\rho}_Q = x'_Q \hat{i}' + y'_Q \hat{j}' + z'_Q \hat{k}', \quad \bar{r}_Q = x_Q \hat{i} + y_Q \hat{j} + z_Q \hat{k} \tag{3.2}$$

Referring to Fig. 3.2, vectors $\bar{\rho}'_Q$ and $\bar{r}_Q$ have the following relationship:

$$\{\bar{r}_Q\} = \{r_o\} + [T]^T \{\rho'_Q\} \tag{3.3}$$

where [T] is a 3×3 orthogonal transformation matrix. The angular orientation of the (x', y', z') system with respect to the (x, y, z) system is specified by the nine components of the [T] matrix and can be written as a function of the three variables, θ, ϕ, and ψ:

$$T = T(\theta, \phi, \psi) \tag{3.4}$$

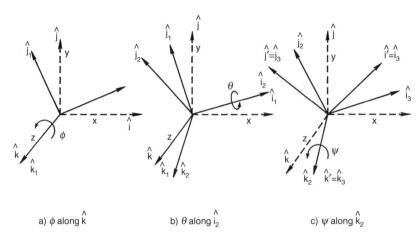

a) ϕ along $\hat{k}$ b) θ along $\hat{i}_2$ c) ψ along $\hat{k}_2$

FIGURE 3.3   Successive rotations of θ, ϕ, and ψ of the (x, y, z) coordinate system.

There are several systems of variables such as θ, ϕ, and ψ which can be used to specify T. At the present formulation the Euler angles are chosen.

The orientation of the moving coordinate system ($\hat{i}'$, $\hat{j}'$ $\hat{k}'$) is obtained from the fixed coordinate system ($\hat{i}$, $\hat{j}$, $\hat{k}$) by applying successive rotation angles, θ, ϕ, and ψ (Fig. 3.3). First, the ($\hat{i}$, $\hat{j}$, $\hat{k}$) system is rotated through an angle ϕ about the z axis (Fig. 3.3a), which results in the intermediary system ($\hat{i}_1$, $\hat{j}_1$ $\hat{k}_1$). The second rotation through an angle θ about the i_1 axis (Fig. 3.3b), produces the intermediary system ($\hat{i}_2$, $\hat{j}_2$ $\hat{k}_2$), and the third rotation through an angle ψ about the k_2 axis (Fig. 3.3c), gives the final orientation of the moving ($\hat{i}'$, $\hat{j}'$ $\hat{k}'$) system relative to the ($\hat{i}$, $\hat{j}$, $\hat{k}$) system. The orthogonal transformation matrix [T] resulting from the above rotations is given *by*

$$[T] = \begin{bmatrix} \cos\phi\cos\psi & \cos\psi\sin\phi & \sin\theta\sin\psi \\ -\sin\psi\cos\theta\sin\phi & +\sin\psi\cos\theta\cos\phi & \\ -\sin\psi\cos\phi & -\sin\psi\sin\phi & \cos\psi\sin\theta \\ -\cos\psi\cos\theta\sin\phi & +\cos\psi\cos\theta\cos\phi & \\ \sin\phi\sin\theta & -\sin\theta\cos\phi & \cos\theta \end{bmatrix} \qquad (3.5)$$

Joint Surfaces and Contact Conditions

Assuming rigid body contacts between the two body segments at points C_i ($i = 1, 2$) as shown in Fig. 3.2, let us represent the contact surfaces by smooth mathematical functions of the following form:

$$y = f(x, z), \quad y' = g(x', z') \qquad (3.6)$$

As implied in Eq. (3.6), y and y' represent the fixed and the moving surfaces, respectively. The position vectors of the contact points C_i ($i = 1, 2$) in the base ($\hat{i}$, $\hat{j}$, $\hat{k}$) are denoted by

$$\bar{r}_{c_i} = x_{c_i}\hat{i} + f\left(x_{c_i}, z_{c_i}\right)\hat{j} + z_{c_i}\hat{k} \qquad (3.7)$$

and the corresponding ones in the base ($\hat{i}'$, $\hat{j}'$ $\hat{k}'$) are given by

$$\bar{\rho}'_{c_i} = x'_{c_i}\hat{i}' + g\left(x'_{c_i}, z'_{c_i}\right)\hat{j} + z'_{c_i}\hat{k}' \qquad (3.8)$$

Then, at each contact point C_i, the following relationship must hold:

$$\{r_{c_i}\} = \{r_o\} + [T]^T\{\rho'_{c_i}\}$$ (3.9)

This is a part of the geometric compatibility condition for the two contact surfaces. Furthermore, the unit normals to the surfaces of the moving and fixed body segments at the points of contacts must be colinear.

Let $\bar{n}_{c_i}$ ($i = 1, 2$) be the unit normals to the fixed surface, $y = f(x, z)$, at the contact points, $C_i(i = 1, 2)$, then

$$\hat{n}_{c_i} = \frac{1}{\sqrt{\det[G]}}\left(\frac{\partial \bar{r}_{c_i}}{\partial x_{c_i}}\right) \times \left(\frac{\partial \bar{r}_{c_i}}{\partial z_{c_i}}\right) \quad i = 1, 2$$ (3.10)

where $\bar{r}_{c_i}$ is given in Eq. 3.7 and the components of the matrix $[G]$ are determined by

$$G_{kl} = \left(\frac{\partial \bar{r}_{c_i}}{\partial x^k} \cdot \frac{\partial \bar{r}_{c_i}}{\partial x^l}\right) \quad i = 1, 2$$ (3.11)

with $x^1 = x_{c_i}$, $x^2 = (x_{c_i}, z_{c_i})$ $x^3 = f(x_{c_i}, z_{c_i})$.

Therefore, the components of matrix $[G]$ may be expressed as

$$G_{xx} = \left(\frac{\partial x_{c_i}}{\partial x_{c_i}}\right)^2 + \left(\frac{\partial y_{c_i}}{\partial x_{c_i}}\right)^2 + \left(\frac{\partial z_{c_i}}{\partial x_{c_i}}\right)^2$$ (3.12a)

$$G_{zz} = \left(\frac{\partial x_{c_i}}{\partial x_{c_i}}\right)^2 + \left(\frac{\partial y_{c_i}}{\partial z_{c_i}}\right)^2 + \left(\frac{\partial z_{c_i}}{\partial z_{c_i}}\right)^2$$ (3.12b)

$$G_{xz} = G_{zx} = \left(\frac{\partial x_{c_i}}{\partial x_{c_i}}\right)\left(\frac{\partial x_{c_i}}{\partial z_{c_i}}\right) + \left(\frac{\partial y_{c_i}}{\partial x_{c_i}}\right)\left(\frac{\partial y_{c_i}}{\partial z_{c_i}}\right) + \left(\frac{\partial z_{c_i}}{\partial x_{c_i}}\right)\left(\frac{\partial z_{c_i}}{\partial z_{c_i}}\right)$$ (3.12c)

Since $(\partial z_{c_i}/\partial x_{c_i}) = 0$ and $(\partial x_{c_i}/\partial z_{c_i}) = 0$, then the components of matrix $[G_{k\ell}]$ reduce to

$$G_{xx} = 1 + \left(\frac{\partial f}{\partial x_{c_i}}\right)^2$$ (3.13a)

$$G_{zz} = 1 + \left(\frac{\partial f}{\partial z_{c_i}}\right)^2$$ (3.13b)

$$G_{xz} = G_{zx} = \left(\frac{\partial f}{\partial x_{c_i}}\right)\left(\frac{\partial f}{\partial z_{c_i}}\right)$$ (3.13c)

From Eqs. 13a through 13c, the det $[G]$ can be written

$$\det[G] = 1 + \left(\frac{\partial f}{\partial x_{c_i}}\right)^2 + \left(\frac{\partial f}{\partial z_{c_i}}\right)^2$$ (3.14)

and therefore the unit outward normals expressed in Eq. 3.10 will have the following form:

$$\hat{n}_{c_i} = \frac{\gamma}{\sqrt{1 + \left(\dfrac{\partial f}{\partial x_{c_i}}\right)^2 + \left(\dfrac{\partial f}{\partial z_{c_i}}\right)^2}} \left[\left(\frac{\partial f}{\partial x_{c_i}}\right)\hat{i} - \hat{j} + \left(\frac{\partial f}{\partial z_{c_i}}\right)\hat{k}\right] \tag{3.15}$$

where the parameter γ is chosen such that $\hat{n}_{c_i}$ represents the outward normal. Similarly, following the same procedure as outlined above $\hat{n}'_{c_i}$ ($i = 1, 2$), the unit outward normal to the moving surface, $y' = g(x', z')$, at contact points, C_i ($i = 1, 2$), and expressed in the $\hat{x}', \hat{y}', \hat{z}'$ system, can be written as

$$\hat{n}'_{c_i} = \frac{\beta}{\sqrt{1 + \left(\dfrac{\partial g}{\partial x'_{c_i}}\right)^2 + \left(\dfrac{\partial g}{\partial z'_{c_i}}\right)^2}} \left[\left(\frac{\partial g}{\partial x'_{c_i}}\right)\hat{i}' - \hat{j}' + \left(\frac{\partial g}{\partial z'_{c_i}}\right)\hat{k}'\right] \tag{3.16}$$

where the parameter β is chosen such that $\hat{n}'_{c_i}$ represents the outward normal. Colinearity of unit normals at each contact point C_i ($i = 1, 2$), requires that

$$\left\{n_{c_i}\right\} = [T]^T \left\{\hat{n}'_{c_i}\right\} \tag{3.17}$$

Note that colinearity condition can also be satisfied by requiring that the cross product

$$\left(\hat{n}'_{c_i} \times T^T \hat{n}'_{c_i}\right)$$

be zero.

Ligament and Contact Forces

During its motion the moving body segment is subjected to the ligament forces, contact forces, and externally applied forces and moments (Fig. 3.4). The contact forces and the ligament forces are the unknowns of the problem and the external forces and moments will be specified. These forces are discussed in some detail in the following paragraphs.

The ligaments are modeled as nonlinear elastic springs. To be more specific, for the major ligaments of the knee joint, the following force-elongation relationship can be assumed:

$$F_j = K_j(L_j - \ell_j)^2 \text{ for } L_j > \ell_j \tag{3.18a}$$

in which K_j is the spring constant, L_j and ℓ_j are, respectively, the current and initial lengths of the ligament j. The tensile force in the jth ligament is designated by F_j. It is assumed that the ligaments cannot carry any compressive force; accordingly,

$$F_j = 0 \text{ for } L_j < \ell_j \tag{3.18b}$$

The stiffness values, K_j, are estimated according to the data available in the literature.[24,36]

Let $\left(\bar{\rho}'_j\right)_m$ be the position vector in the base $\hat{i}', \hat{j}' \hat{k}'$ of the insertion point of the ligament j in the moving body segment. The position vector of the origin point of the same ligament, in the fixed body

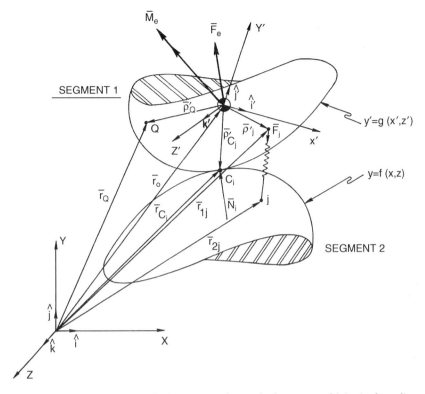

FIGURE 3.4 Forces acting on the moving body segment of a two-body segmented joint in three dimensions.

segment is denoted by $\left(\bar{r}_{2j}\right)_f$ in the base $\hat{i}, \hat{j}, \hat{k}$. The subscripts m and f outside the parenthesis imply "moving" and "fixed," respectively. The current length of the ligament is given by

$$L_j = \sqrt{\left[\left(\bar{r}_{2j}\right)_f - \bar{r}_0 - T^T\left(\bar{\rho}'_j\right)_m\right] \cdot \left[\left(\bar{r}_{2j}\right)_f - \bar{r}_0 - T^T\left(\bar{\rho}'_j\right)_m\right]} \tag{3.19}$$

The unit vector, $\hat{\lambda}_j$, along the ligament j directed from the moving to the fixed body segment is

$$\hat{\lambda}_j = \frac{1}{L_j}\left[\left(\bar{r}_{2j}\right)_f - \bar{r}_0 - T^T\left(\bar{\rho}'_j\right)_m\right] \tag{3.20}$$

Thus, the axial force in the ligament j in its vectorial form becomes

$$\bar{F}_j = F_j\hat{\lambda}_j \tag{3.21}$$

where F_j is given by Eq. 3.18. Since the friction force between the moving and fixed body segment is neglected, the contact force will be in the direction of the normal to the surface at the point of contact. The contact forces, N_i, acting on the moving body segment are given by

$$\bar{N}_i = \left|N_i\right|\left[\left(n_{c_i}\right)_x \hat{i} + \left(n_{c_i}\right)_y \hat{j} + \left(n_{c_i}\right)_z \hat{k}\right] \tag{3.22}$$

where $\left|N_i\right|$ represents the unknown magnitudes of the contact forces and $(n_{ci})_x$, $(n_{ci})_y$, and $(n_{ci})_z$ are the components of the unit normal, n_{ci}, in the x, y, and z directions, respectively.

In general, the moving body segment of the joint is subjected to various external forces and moments whose resultants at the center of mass of the moving body segment are given as

$$\overline{F}_e = \left(F_e\right)_x \hat{i} + \left(F_e\right)_y \hat{j} + \left(F_e\right)_z \hat{k}, \quad \overline{M}_e = \left(M_e\right)_x \hat{i} + \left(M_e\right)_y \hat{j} + \left(M_e\right)_z \hat{k} \tag{3.23}$$

Equations of Motion

The equations governing the forced motion of the moving body segment are

$$\left(F_e\right)_x + \sum_{i=1}^{q} \left|N_i\right| \left(n_{c_i}\right)_x + \sum_{j=1}^{p} F_j \left(\lambda_j\right)_x = M\ddot{x}_o \tag{3.24a}$$

$$\left(F_e\right)_y + \sum_{i=1}^{q} \left|N_i\right| \left(n_{c_i}\right)_y + \sum_{j=1}^{p} F_j \left(\lambda_j\right)_x = M\ddot{y}_o \tag{3.24b}$$

$$\left(F_e\right)_z + \sum_{i=1}^{q} \left|N_i\right| \left(n_{c_i}\right)_z + \sum_{j=1}^{p} F_j \left(\lambda_j\right)_z = M\ddot{z}_o \tag{3.24c}$$

$$\sum M_{x'x'} = I_{x'x'}\dot{\omega}_{x'} + \left(I_{z'z'} - I_{y'y'}\right)\omega_{y'}\omega_{z'} \tag{3.25a}$$

$$\sum M_{y'y'} = I_{y'y'}\dot{\omega}_{y'} + \left(I_{x'x'} - I_{z'z'}\right)\omega_{z'}\omega_{x'} \tag{3.25b}$$

$$\sum M_{z'z'} = I_{z'z'}\dot{\omega}_{z'} + \left(I_{y'y'} - I_{x'x'}\right)\omega_{x'}\omega_{y'} \tag{3.25c}$$

where p and q represent the number of ligaments and the contact points, respectively. $I_{x'x'}$, $I_{y'y'}$, and $I_{z'z'}$ are the principal moments of inertia of the moving body segment about its centroidal principal axis system (x', y', z'); and $\omega_{x'}$, $\omega_{y'}$, and $\omega_{z'}$ are the components of the angular velocity vector which are given below in terms of the Euler angles:

$$\omega_{x'} = \dot{\theta}\cos\psi + \dot{\phi}\sin\theta\sin\psi, \quad \omega_{y'} = -\dot{\theta}\sin\psi + \dot{\phi}\sin\theta\cos\psi, \quad \omega_{z'} = \dot{\phi}\cos\theta + \dot{\psi} \tag{3.26}$$

The angular acceleration components, $\dot{\omega}_x$, $\dot{\omega}_y$, and $\dot{\omega}_z$ are directly obtained from Eq. 3.26:

$$\dot{\omega}_{x'} = \ddot{\theta}\cos\psi - \dot{\psi}\left(\dot{\theta}\sin\psi - \dot{\phi}\cos\psi\sin\theta\right) + \ddot{\phi}\sin\theta\sin\psi + \dot{\phi}\dot{\theta}\cos\theta\sin\psi \tag{3.27a}$$

$$\dot{\omega}_{y'} = -\ddot{\theta}\sin\psi - \dot{\psi}\left(\dot{\theta}\cos\psi - \dot{\phi}\sin\psi\sin\theta\right) + \ddot{\phi}\sin\theta\cos\psi + \dot{\phi}\dot{\theta}\cos\theta\sin\psi \tag{3.27b}$$

$$\dot{\omega}_{z'} = \ddot{\phi}\cos - \dot{\phi}\dot{\theta}\sin\theta + \ddot{\psi} \tag{3.27c}$$

Note that the moment components shown on the left-hand sides of Eqs. 3.25a through 3.25c have the following terms:

$$\overline{M} = \overline{M}_e + \sum_{i=1}^{q} [T](\overline{\rho}'_{c_i}) \times (|N_i|\overline{n}_{c_i}) + \sum_{j=1}^{p} [T](\overline{\rho}'_{c_i}) \times (F_j\overline{\lambda}_j) \tag{3.28}$$

where $\overline{M}_e$ is the applied external moment, and p and q again represent the number of ligaments and contact points, respectively.

Eqs 3.24 and 3.25 form a set of six nonlinear second-order differential equations which, together with the contact conditions (3.9) and (3.17), form a set of 16 nonlinear equations, assuming two contact points, i.e., (i = 1, 2) with 16 unknowns:

(1) θ, ϕ, and ψ, which determine the components of transformation matrix $[T]$
(2) x_o, y_o, and z_o: the components of position vector $\overline{r}_o$
(3) x_{ci}, z_{ci}, x'_{ci} and x'_{ci} (i = 1, 2): the coordinates of contact points
(4) $|N_i|$ (i = 1, 2): the magnitudes of the contact forces

The problem description is completed by assigning the initial conditions which are

$$\dot{x}_o = \dot{y}_o = \dot{z}_o = 0, \quad \dot{\omega}_x = \dot{\omega}_y = \dot{\omega}_z = 0 \tag{3.29}$$

along with specified values for x_o, y_o, z_o, θ, ϕ, and ψ, at t = 0. Before we describe the numerical procedure employed in the solution of the governing equations in the next section, the following observation must be made. During its motion, segment 1 is subjected to the ligament forces, contact forces, and the externally applied forces and moments, Fig. 3.4. The contact force and the ligament forces are the unknowns of the problem and the external forces and moments are specified. The reader might wonder why the muscle forces are not included in the dynamic modeling of an articulating joint. If the model under consideration is intended to simulate events which take place during a very short time period such as 0.1 seconds, then it is sufficient to consider only the passive resistive forces at the model formulation. However, direct exclusion of the muscle forces from the model does not restrict its capabilities to have the effects of muscle forces included, if desired, as a part of the applied force and moment vector on the moving body segment.

Numerical Solution Procedure

The governing equations of the initial value problem at hand are the six equations of motion (3.24) and (3.25), four contact conditions (3.9) and six geometric compatibility conditions (3.17). The main unknowns of the problem are x_o, y_o, z_o, θ, ϕ, ψ, x_{c1}, z_{c1}, x_{c2}, z_{c2}, x'_{c1}, z'_{c1}, x'_{c2}, z'_{c2}, N_1, and N_2. The problem is thus reduced to the solution of a set of simultaneous nonlinear differential and algebraic equations.

The first step in arriving at a numerical solution of these equations is the replacement of the time derivatives with temporal operators; in the present work, the Newmark operators[2] are chosen for this purpose. For instance, x_o is expressed in the following form:

$$\ddot{x}_o^t = \frac{4}{(\Delta t)^2}(x_o^t - x_o^{t-\Delta t}) - \frac{4}{\Delta t}\dot{x}_o^{t-\Delta t} - \ddot{x}_o^{t-\Delta t}, \quad \dot{x}_o^t = \dot{x}_o^{t-\Delta t} + \frac{\Delta t}{2}\ddot{x}_o^{t-\Delta t} + \frac{\Delta t}{2}\ddot{x}_o^t \tag{3.30}$$

in which Δt is the time increment and the superscripts refer to the time stations. Similar expressions may be used for $\ddot{y}_o$, $\ddot{z}_o$, $\ddot{\phi}$, $\ddot{\theta}$, and $\ddot{\psi}$. In the applications of Eq. 3.30, the conditions at the previous time station (t = Δt) are, of course, assumed to be known.

After the time derivatives in Eqs 3.24 and 3.25 are replaced with the temporal operators defined, the governing equations take the form of a set of nonlinear algebraic equations. The solution of these equations is complicated by the fact that iteration or perturbation methods must be used. In this work, the Newton-Raphson[23] iteration is used for the solution. To linearize the resulting set of simultaneous algebraic equations we assume

$$^{k}x_{o}^{t} = {}^{k\text{-}1}x_{o}^{t} + \Delta x_{o} \tag{3.31}$$

and similar expressions for the other variables are written. Here, the right superscripts denote the time station under consideration and the left superscripts denote the iteration number. At each iteration k the values of the variables at the previous $(k - 1)$ iteration are assumed to be known. The delta quantities denote incremental values. Eq. 3.31 and those corresponding to the other variables are substituted into the governing nonlinear algebraic equations and the higher order terms in the delta quantities are dropped. The set of n simultaneous algebraic (now linearized) equations can be put into the following matrix form:

$$[K]\{\Delta\} = \{D\} \tag{3.32}$$

where [K] is an n × n coefficient matrix, $\{\Delta\}$ is a vector of incremental quantities, and $\{D\}$ is a vector of known values.

The iteration process at a fixed time station continues until the delta quantities of all the variables become negligibly small. A solution is accepted and the iteration process is terminated when the delta quantities become less than or equal to a small increment of the previous values of the corresponding variables. The converged solution of each variable is then used as the initial value for the next time step and the process is repeated for consecutive time steps. The only problem that the Newton-Raphson process may present in the solution of dynamic problems is due to the fact that the period of the forced motion of the system may turn out to be quite short. In this case it becomes necessary to use very small time steps; otherwise, a significantly large number of iterations is required for convergence.

Numerical procedures outlined above can be utilized, in principle, for the solution of the three-dimensional joint model equations presented so far. However, because of the extreme complexity of these equations, in the next section we will only present some representative numerical results of the two-dimensional version of our formulation applied to the dynamic model of the human knee joint.

3.3 Two-Dimensional Dynamic Joint Model and Various Solution Methods

Model Description

The two-dimensional version of the dynamic human knee model introduced by Engin and Moeinzadeh[7,11,13,14] involves two body segments representing femur and tibia which execute a relative dynamic motion in the sagittal plane within the constraints of ligaments as shown schematically in Fig. 3.5. An inertial coordinate system (x, y) is connected to the femur which is assumed to be fixed, while the moving coordinate system (u, v) is attached to the center of mass of the lower leg. The coordinates x_0 and y_0 of the origin of the moving (u, v) system and its orientation θ with respect to the (x, y) system define the motion of the lower leg relative to the upper leg. Articulating surfaces are represented by a fourth order polynomial $y = f(x)$ for the femur, and a second order polynomial $v = g(u)$ for the tibia.

The following three differential equations describe the forced motion of the lower leg relative to the femur:

$$R_{x} + Ne_{nx} + \sum_{j=1}^{4} F_{jx} = m\ddot{x}_{0} \tag{3.33}$$

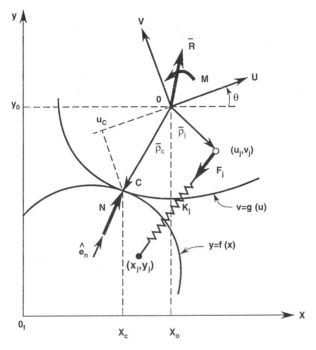

FIGURE 3.5   Schematic drawing of the two-body segmented joint model. (*Source*: Engin, A.E. and Tumer, S.T., *J. Biomechan. Eng.*, ASME, 1993. With permission.)

$$R_y + Ne_{ny} + \sum_{j=1}^{4} F_{jy} = m\ddot{y}_0 \tag{3.34}$$

$$M + N\left(\rho_{c_x} e_{ny} - \rho_{c_y} e_{nx}\right) + \sum_{j=1}^{4}\left(\rho_{jx} F_{jy} - \rho_{jy} F_{jx}\right) = I\ddot{\theta} \tag{3.35}$$

where, R_x, R_y, and M are the results of all forces, including muscle actions and externally applied transient forces, acting on the lower leg at its mass center; N is the contact force between the femur and tibia; F_j represents the forces in the four major ligaments of the knee joint; m is the mass; and I is the centroidal mass moment of inertia of the lower leg. The subscripts x and y denote the x and y components, respectively, of the vectors $\hat{e}_n$, $\bar{\rho}_c$, $\bar{\rho}_j$ and $\bar{F}_j$ all defined in Fig. 3.5.

Designating the x and u coordinates of contact point C by x_c and u_c the following geometric equations, which express the contact conditions, are written:

$$x_c = x_0 + u_c \cos\theta - g(u_c)\sin\theta \tag{3.36}$$

$$f(x_c) = y_0 + u_c \sin\theta + g(u_c)\cos\theta \tag{3.37}$$

$$\sin\theta[1 + f'(x_c)\,g'(u_c)] - \cos\theta[f'(x_c) - g'(u_c)] = 0 \tag{3.38}$$

The rationale behind the contact conditions and the derivations of these equations are provided in Section 3.2. Briefly, Eqs. 3.36 and 3.37 represent the geometric compatibility and Eq. 3.38 is the condition of colinearity of normals at the point of contact. Note that Eq. 3.33 needs to be solved for six unknowns, x_0, y_0, θ, x_C, u_C, and N, for given initial conditions and forcing.

The four major ligaments, namely, the anterior cruciate (AC), posterior cruciate (PC), medial collateral (MC), and lateral collateral (LC) are treated as nonlinear springs working only in tension. In the original model,[11] quadratic force-elongation characteristics were considered for the ligament behavior. The method requires the initial lengths to be known. It was assumed that all ligaments were relaxed at a flexion angle of 55° and simulation was started from this configuration. Alternatively,[16] ligament behavior is modified so that the need to specify an initial configuration where all ligaments are simultaneously at their relaxed state is no longer necessary. Instead, the available data on the strain levels of ligaments at full extension of the knee are used.[39] Accordingly, the following constitutive relation with parabolic and linear regions is adopted for a particular ligament *j*:

$$F_j = \frac{1}{4} k_j \frac{\varepsilon_j^2}{\varepsilon_b} \quad \text{for} \quad 0 \le \varepsilon_j \le 2\varepsilon_b \tag{3.39a}$$

$$F_j = k_j(\varepsilon_j - \varepsilon_b) \text{ for } \varepsilon_j > 2\varepsilon_b \tag{3.39b}$$

where, k_j is the spring constant in N per unit strain, ε_j is the current strain, and $2\varepsilon_b$ is the strain level at the boundary between the parabolic and linear portions of the force-strain relation. The value suggested in Reference 39 for ε_b is 3% for all ligaments. The strain ε_j of ligament *j* is given by

$$\varepsilon_j = \frac{l_j\left(1 + \varepsilon_{je}\right) - l_{je}}{l_{je}} \tag{3.39c}$$

where l_j is the current length, ε_{je} is the strain at full extension, and l_{je} is the length at full extension. Knowing the origins (x_j, y_j) and insertions (u_j, v_j) of ligaments, their lengths at any desired knee configuration can be calculated.

Furthermore, coordinates of insertions and origins of the cruciate ligaments are modified by monitoring the way in which they cross each other and the relation of the crossing point to the tibio-femoral contact point in accordance with previously established observations.[19] This modification is accomplished with the help of an interactive animation program which displays all four ligaments and articulating surfaces on the screen at successive knee configurations during the course of relative dynamic motion.

Detailed discussions of various anatomical and functional aspects of the human knee joint can be found in References 8 and 12. The first task in obtaining numerical results is determination of the functions $f(x)$ and $g(u)$ from an X-ray of a human knee joint. A number of points on the two-dimensional profiles of the femoral and tibial articulating surfaces are utilized to obtain quartic and quadratic polynomials, respectively.

Two types of external forces which pass through the center of mass of the tibia and perpendicular to the long axis of the tibia are considered. The first one is a rectangular pulse of duration t_o and the second one is an exponentially decaying sinusoidal pulse of the same duration. A dynamic loading of the form of a rectangular pulse is extremely difficult to simulate experimentally. Exponentially decaying sinusoidal pulse has been previously used[18] as a typical dynamic load in head impact analysis. The effect of pulse duration on the dynamic response of the knee joint is examined by taking t_o equal to 0.05, 0.10, and 0.15 seconds for both rectangular and exponentially decaying sinusoidal pulses. The effect of pulse amplitude, A, is also investigated by taking A equal to 20, 60, 100, 140, and 180 Newtons for both types of pulses. Some representative results obtained by means of the numerical solution technique outlined in the previous section are presented in Figs. 3.6 through 3.8. The values in parenthesis in Figs. 3.7 and 3.8 indicate the knee flexion angles at the corresponding times. Several remarks can be made about the ligament and contact forces. When the knee joint is extended dynamically, all major ligaments with the exception of the posterior ligament are elongated. The magnitudes of the anterior cruciate ligament forces and the corresponding contact forces in response to a particular forcing function are comparable as depicted in Figs. 3.7 and 3.8.

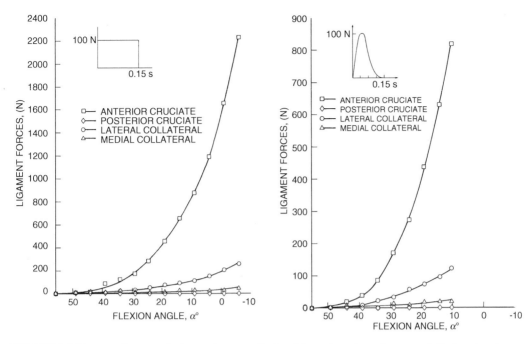

FIGURES 3.6 Ligament forces vs. flexion angle, α, are displayed for rectangular and exponentially decaying sinusoidal pulses of 0.15-second durations.

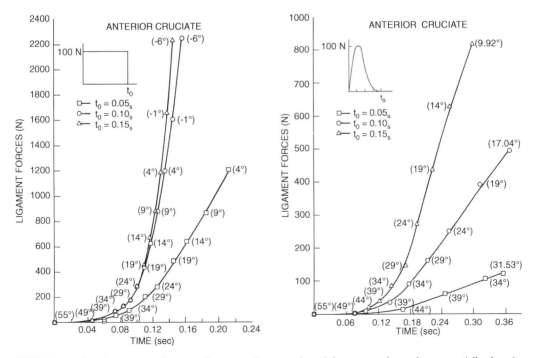

FIGURES 3.7 Anterior cruciate ligament forces vs. time are plotted for rectangular and exponentially decaying sinusoidal pulses of various durations.

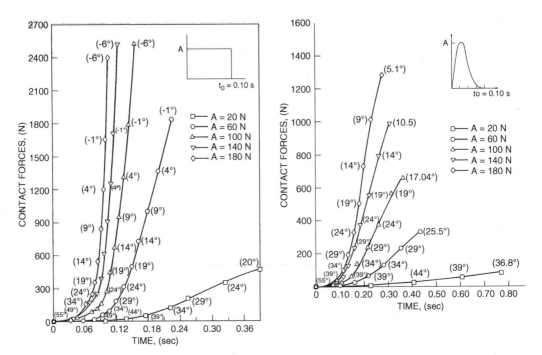

FIGURES 3.8 Joint contact forces vs. time are shown for rectangular and exponentially decaying sinusoidal pulses of various amplitudes.

Alternative Methods of Solution

The initial value problem described in the previous section consists of three nonlinear differential equations (3.33 through 3.35) coupled with three nonlinear constraint equations (3.36 through 3.38). Replacing the time derivatives in the differential equations by a temporal operator and solving the resulting set of algebraic equations by iteration at every fixed time station constitute the previous method of solution. Along the lines already utilized in dynamics of multidegree-of-freedom mechanical systems,[29] a completely different approach in which the constraint equations are differentiated twice and the resulting second-order simultaneous differential equations are numerically integrated is proposed as a first alternative. The basic postulate of this approach is that if the constraints are satisfied initially, then satisfying the second derivatives of the constraints in future time steps would also satisfy the constraints themselves. This method is called the method of excess differential equations (EDEs), and its application to the problem at hand is outlined below.

Upon differentiating the constraint equations (3.36 through 3.38) twice, we now have, by including the original differential equations (3.33 through 3.35), a set of six coupled second-order differential equations which can be arranged into the following form:

$$[A]\left[\ddot{x}_0 \ddot{y}_0 \ddot{\theta} \, \ddot{x}_c \ddot{u}_c N\right]^T = \left[F_1 \ldots, F_6\right]^T \tag{3.40}$$

where, $[A]$ is a 6×6 configuration-dependent coefficient matrix, $[F_1, \ldots, F_6]^T$ is a configuration and time-dependent forcing vector, and the superscript T stands for transpose. Solving for the unknown vector of accelerations and contact force we can get

$$\left[\ddot{x}_0 \ddot{y}_0 \ddot{\theta} \, \ddot{x}_c \ddot{u}_c\right]^T = \left[S_1, \ldots, S_5\right]^T \quad \text{and} \quad N = S_6 \tag{3.41}$$

where, S_1, etc. are the elements of the vector $[A]^{-1}[F]^T$, and are expressed in terms of five position variables $(x_0, y_0, \theta, x_C, u_C)$, their first derivatives, and the known external forces at any time. By knowing the position variables and their first derivatives at the previous time station, the first part of Eq. 3.41 can be numerically integrated to find position variables and their first derivatives at the current time. The corresponding contact force can then be found from the second part of Eq. 3.41. The integration process can be repeated as many times as required until the total time of simulation is reached. Note that this method involves far less mathematical manipulation than the previous iteration method, and more importantly, numerical solution is restricted to the integration process which does not require iteration.

Ideally, one would like to have a minimum number of simultaneous differential equations describing the dynamics of a system. Since the biomechanical system at hand has two rigid body degrees-of-freedom, its dynamics can, in principle, be expressed by two differential equations in terms of two appropriately chosen generalized coordinates. For the present human knee model, θ and x_C are chosen as the generalized coordinates. This approach, called the method of minimal differential equations (MDE),[17,29] is introduced as a second alternative to the original solution technique.

Since the constraint equations (3.36 through 3.38) are linear in terms of velocity variables, it is possible to express $\dot{x}_0$ and $\dot{y}_0$ as linear combinations of generalized velocities:

$$\dot{x}_0 = \lambda_\theta \dot{\theta} + \lambda_x \dot{x}_c \quad \text{and} \quad \dot{y}_0 = \mu_\theta \dot{\theta} + \mu_x \dot{x}_c \tag{3.42}$$

Components of mass center acceleration can then be expressed as:

$$\ddot{x}_0 = \lambda_\theta \ddot{\theta} + \lambda_x \ddot{x}_c + \lambda_d \quad \text{and} \quad \ddot{y}_0 = \mu_\theta \ddot{\theta} + \mu_x \ddot{x}_c + \mu_d \tag{3.43}$$

where the expressions for $\lambda\theta$, λ_x, λ_d, $\mu\theta$, λ_x, and λ_d are given below:

$$\lambda_\theta = \left[g(u_c)\cos\theta + u_c\sin\theta \right] + \left[\frac{g'(u_c)\left[\sin\theta - f'(x_c)\cos\theta\right]}{g''(u_c)\left[\sin\theta - f'(x_c)\cos\theta\right]} - \frac{1}{g''(u_c)} \right]\left[g'(u_c)\sin\theta - \cos\theta\right]$$

$$\lambda_x = 1 + \left[\frac{f''(x_c)\left[\cos\theta - u'(x_c)\sin\theta\right]}{g''(u_c)\left[\cos\theta + f'(x_c)\sin\theta\right]} \right]\left[g'(u_c)\sin\theta - \cos\theta\right]$$

$$\mu_\theta = \left[g(u_c)\sin\theta - u_c\cos\theta \right] - \left[\frac{g'(u_c)\left[\sin\theta - f'(x_c)\cos\theta\right]}{g''(u_c)\left[\cos\theta - f'(x_c)\sin\theta\right]} - \frac{1}{g''(u_c)} \right]\left[\sin\theta + g'(u_c)\cos\theta\right]$$

$$\mu_x = f'(x_c) - \frac{f''(x_c)\left[\cos\theta - g'(x_c)\sin\theta\right]}{g''(u_c)\left[\cos\theta + f'(x_c)\sin\theta\right]}\left[\sin\theta + g'(u_c)\cos\theta\right]$$

$$\lambda_d = \frac{\partial\lambda_\theta}{\partial\theta}\dot{\theta}^2 + \frac{\partial\lambda_x}{\partial x_c}\dot{x}_c^2 + \left(\frac{\partial\lambda_\theta}{\partial x_c} + \frac{\partial\lambda_x}{\partial\theta}\right)\dot{\theta}\dot{x}_c$$

$$\mu_d = \frac{\partial\mu_\theta}{\partial\theta}\dot{\theta}^2 + \frac{\partial\mu_x}{\partial x_c}\dot{x}_c^2 + \left(\frac{\partial\mu_\theta}{\partial x_c} + \frac{\partial\mu_x}{\partial\theta}\right)\dot{\theta}\dot{x}_c$$

We can arrange the original differential equations (3.38 through 3.35) into the following form by using elements of matrix $[A]$ and vector $[F]^T$ given in Eq. 3.40:

$$a_{11}\ddot{x}_0 + a_{16}N = F_1 \tag{3.44}$$

$$a_{22}\ddot{y}_0 + a_{26}N = F_2 \tag{3.45}$$

$$a_{33}\ddot{\theta} + a_{36}N = F_3 \tag{3.46}$$

We then solve for N from Eq. 3.46 and substitute into Eqs. 3.44 and 3.45 together with Eq. 3.43, and thus obtain

$$[B]\left[\ddot{\theta}\,\ddot{x}_c\right]^T = \left[H_1 H_2\right]^T \tag{3.47}$$

where, $[B]$ is a 2×2 configuration-dependent coefficient matrix, and $[H]^T$ is a configuration- and time-dependent forcing vector. Eq. 3.47 can now be integrated to obtain the dynamic response in terms of generalized coordinates $\theta(t)$ and $x_c(t)$. The contact force, N, is directly found from Eq. 3.46. It is necessary to solve the geometric constraint equations after every integration step in order to carry on with the next step. The nature of the constraint equations (3.36 through 3.33) allows one to obtain closed form expressions for x_0, y_0, and u_C in terms of generalized coordinates θ and x_C.

As one might expect, these two methods are mathematically equivalent. In fact, after a series of row operations on matrix Eq. 3.40, one can show that Eq. 3.47 is a partitioned form of Eq. 3.40. However, from a numerical solution point of view, these methods are not equivalent. In the MDE method, the constraints are directly satisfied at every integration step, whereas, in the EDE method, constraints are directly satisfied only at the initial time. On the other hand, EDE formulation is quite straightforward and can be readily applied to any problem of this kind. The MDE method requires a proper choice of generalized coordinates in the first place; even then it might not always be possible to arrive at the desired formulation which does not involve iteration.

Both the excess and minimal differential equations methods have been programmed in Quick Basic by utilizing two different integration schemes for the two-dimensional model of the human knee. The Euler method constitutes the crudest numerical integration method, whereas the fourth-order Runge-Kutta (R-K) algorithm is considered to be a more sophisticated and accurate alternative. The four combinations of two formulations and two methods of integration have been tested by several types of pulses applied to the lower leg. The results are presented in Table 3.1 for a typical pulse for comparison.

Most of the calculations are essentially the same, so formulations of the excess and minimal differential equations take practically the same amount of time. As expected, the Runge-Kutta algorithm requires considerably more time than the Euler integration. Considering the results of the R-K plus MDE combination as the base values, percentage variations in the maximum values of the contact force, force in the anterior cruciate ligament, and the maximum knee extension reached are shown in Table 3.1. The results indicate that all four combinations yield stable solutions with reasonably small variations. Time histories of all the relevant variables showed small variations for the four combinations. Maximum differences are noted to occur at the peak values. However, there are virtually no differences in the times at which peak values occur. Considering the computational cost, the Euler and MDE combination seems to be the best choice. For more complicated problems where the method of minimal differential equations is not feasible, the straightforward application of the method of excess differential equations may prove to be a suitable alternative when used together with a reliable integration scheme.

The results of these methods are also compared with those of the earlier iterative solution of the problem. The maximum deviations are seen to be less than 2%. If one considers the iterative nature of the earlier solution, superiority of the alternative methods may comfortably be claimed for both accuracy

TABLE 3.1   Comparison of MDE and EDE Methods with Euler (Eu) and Runge-Kutta (R-K) Integration Schemes on IBM-PS/ 2

Method	Execution Time (Min:Sec)	Contact Force	Percentage Variation in Maximum Values of	
			A.C. Ligament Force	Knee Extension
R-K + MDE	3:31	–	–	–
Eu + MDE	1:05	0.1	0.1	0.02
R-K + EDE	3:35	0.1	0.1	0.5
Eu + EDE	1:06	2.5	2.1	3.4

(*Source*: Engin, A.E. and Tumer, S.T., *J. Biomechan. Eng.*, ASME, 1993. With permission.)

and efficiency. Furthermore, all shortcomings of the previous iterative method of solution are eliminated by the alternative methods discussed herein. With these improved solution techniques, the dynamic knee model can now be utilized to study the response of the knee to impact loads applied at any location on the lower leg. In the study of impact, one is automatically tempted to apply classical impact theory. It would also be interesting to see to what extent the classical impact theory holds for an anatomically based knee joint model.

3.4   Application of the Impact Theory to Dynamic Joint Models

Classical impact theory is based on the assumption that impact duration is sufficiently short to allow the following simplifications to be made: (1) geometry does not change during impact, and (2) time integrals of finite quantities over duration of impact are negligible. Formulation introduced in the previous section renders relatively straightforward application of the impact theory to the anatomically based model of the human knee joint.

To apply the impact theory to the present model we first integrate equations of motion (3.33 through 3.35) from t = 0 to t = τ, where τ is the impact period. With the above-mentioned assumptions of the impact theory, the equations are simplified and put into the following forms:

$$m \Delta \dot{x}_o + a_{16} S_N = S_x \tag{3.48a}$$

$$m \Delta \dot{y}_o + a_{26} S_N = S_y \tag{3.48b}$$

$$I \Delta \dot{\theta} + a_{36} S_N = H \tag{3.48c}$$

where, Δ indicates change in velocity terms; S_N, S_x, S_y, and H are impulses of N, R_x, R_y, and M, respectively. The coefficients a_{16}, a_{26}, and a_{36} are as defined in Eq. 3.40 in relation to Eqs. 3.33 through 3.35. Upon substituting for $\Delta \dot{x}_0$ and $\Delta \dot{y}_0$ from Eq. 3.42 into Eq. 3.48, we get:

$$\begin{bmatrix} m\lambda_\theta & a_{16} & m\lambda_x \\ m\mu_\theta & a_{26} & m\mu_x \\ I & a_{36} & 0 \end{bmatrix} \begin{Bmatrix} \Delta\dot{\theta} \\ S_N \\ \Delta\dot{x}_C \end{Bmatrix} = \begin{Bmatrix} S_x \\ S_y \\ H \end{Bmatrix} \tag{3.49}$$

Eq. 3.49 can now be solved for the impulse of the contact force, S_N and the change in angular velocity of the lower leg, $\Delta\dot{\theta}$:

$$S_N = \frac{I\left(S_x\mu_x - S_y\lambda_x\right) + mH\left(\lambda_x\mu_\theta - \mu_x\lambda_\theta\right)}{I\left(a_{16}\mu_x - a_{26}\lambda_x\right) + ma_{36}\left(\lambda_x\mu_\theta - \mu_x\lambda_\theta\right)} \qquad (3.50)$$

$$\Delta\dot\theta = \frac{a_{36}\left(S_y\lambda_x - S_x\mu_x\right) + H\left(a_{16}\mu_x - a_{26}\lambda_x\right)}{I\left(a_{16}\mu_x - a_{26}\lambda_x\right) + ma_{36}\left(\lambda_x\mu_\theta - \mu_x\lambda_\theta\right)} \qquad (3.51)$$

In the above equations one can identify the effects of lower leg inertia (m, I), externally applied impulse (S_x, S_y, H), and knee configuration at the time of impact (the remaining terms). It should be noted that the geometric terms include the effect of the form of contact surfaces on the impact phenomenon. Since forces in ligaments are position dependent, according to the impact theory the ligaments cannot sustain any impulse during impact.

Numerical Results and Discussion

Numerical results of the exact (MDE method) and the approximate (impact theory) solutions were obtained by using the coefficients of the articular surface polynomials presented in Engin and Moeinzadeh.[11] The mass and centroidal mass moment of inertia of the leg were taken to be m = 4 kg, I = 0.1 kg m². The results presented here are for an external impact loading applied at a point 0.25 m below the mass center of the lower leg.

Results of the approximate solution are presented in Figs. 3.9 and 3.10. First, an externally applied impulse perpendicular to the tibial axis along posterior direction is considered. Corresponding tibio-femoral contact impulse, normalized with respect to the magnitude of the externally applied impulse, vs. knee flexion angle is plotted in Fig. 3.9. This figure shows a dramatic increase in tibio-femoral contact impulse with increasing knee flexion angle. At the flexion angle of 35°, the influence of the orientation of the external impulse on the normalized contact impulse is given in Fig. 3.10. The fact that maximum contact impulse is obtained at $\beta = 80°$ is a reflection of the effect of knee geometry. If the knee were assumed to be a simple hinge joint, this maximum would have occurred at $\beta = 90°$. It is also observed that while the posteriorly directed external impulse ($\beta = 0°$) gives rise to compressive contact impulse, the anteriorly directed external impulse ($\beta = 180°$) shows the opposite tendency.

In the case of the approximate solution, time profile of the impact loading is equivalent to the Dirac delta function; whereas, in the exact solution, time profile of the impact load can be specified in any desired form. Impact loads have finite durations in physical situations.

We will next present in Figs. 3.11–3.15 results obtained from the application of the exact solution for the following conditions: An impact load of rectangular shape with an impulse of 10 N is applied along the posterior direction on the lower leg at a point 0.25 m below the mass center and perpendicular to the tibial axis ($\beta = 0$). The knee flexion angle is taken to be 35° prior to impact, and two initial conditions are considered for the lower leg. The first case assumes the lower leg to be stationary, and in the second case the lower leg is assumed to have an initial angular velocity of 10 rad/s in the opposite direction to the applied impact load. Figs. 3.11 and 3.12 illustrate the effect of duration of external impulse on the contact impulse and on the magnitude of the maximum contact force, respectively. The result obtained from the approximate solution for the same amount of external impulse is marked in Fig. 3.11.

It is clearly seen that the approximate solution obtained by the application of the classical impact theory is, as expected, a limiting case of the exact solution as impulse duration approaches zero. The difference between the results of contact impulse obtained from both solutions increases dramatically as the impulse duration increases. For a modest impact duration of 10 ms, the difference is larger than 100%. Furthermore, Fig. 3.11 also displays the influence of initial angular velocity of the lower leg, a factor that cannot even be studied with the simplified approximate solution. Contact impulse alone is not sufficient to describe the loading situation at the tibio-femoral articulation. Fig. 3.12 shows the maximum value of the contact force reached during the impulse period. Note that the classical impact

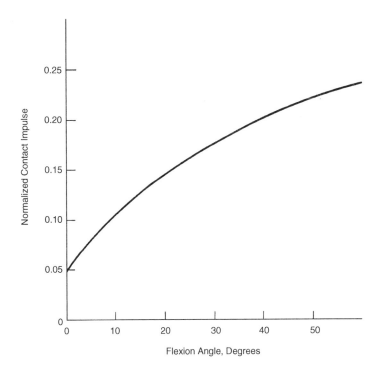

FIGURE 3.9   Contact impulse normalized with respect to external impulse vs. flexion angle.

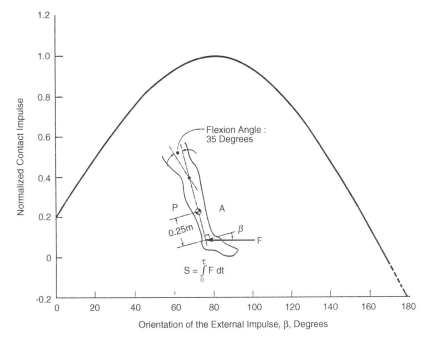

FIGURE 3.10 Contact impulse normalized with respect to external impulse vs. orientation of external impulse (at 35° flexion).

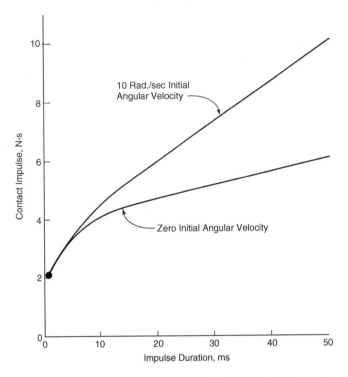

FIGURE 3.11 Contact impulse vs. impulse duration. The result of the approximate solution (classical impact solution) is indicated by (·). (*Source*: Engin, A.E. and Tumer, S.T., *J. Biomechan. Eng.*, ASME, 1993. With permission.)

theory yields an infinite value, and the results of the exact theory approach this limiting value asymptotically as impulse duration approaches zero. Looking at Figs. 3.11 and 3.12 together one can see quite a different trend between the contact impulse and maximum value of the contact force as impulse duration is increased.

Fig. 3.13 shows changes in angular position of the lower leg upon impact as a function of impulse duration. It should be noted that for the limiting case of zero impulse duration there is no change in angular position whether or not the lower leg has an initial velocity. For finite impulse durations and under the conditions prescribed, the knee goes into flexion upon impact when the lower leg is initially stationary, whereas it continues its motion in the extension direction for the case of nonzero initial angular velocity.

The exact solution is also capable of providing information on the time histories of various quantities. Time variations of the contact force and anterior cruciate ligament force are given in Figs. 3.14 and 3.15, respectively, for various impulse durations when the lower leg is stationary prior to impact. Fig. 3.14 indicates sharp increases in contact force levels as duration of external impulse decreases, despite the fact that contact impulse shows the opposite tendency (see Fig. 3.11). Furthermore, although not shown in the figure, after the termination of external impulse, the contact force shows a sudden drop to a value that may be attributed to ligament and inertia forces. In Fig. 3.15, the anterior cruciate force-time curves are drawn beyond the end of the corresponding impulse duration and shown by dashed lines. One may observe that the maximum value of anterior cruciate ligament force increases as the duration of externally applied pulse gets smaller. For small impulse durations, maximum values occur after the external pulse ceases, unlike contact force behavior.

The results presented in this section clearly establish the fact that classical impact theory gives the limiting solution to the model equations as the impact time approaches zero. Moreover, the results indicate *inapplicability* of the classical impact theory to practical situations where the impact time can range from 15 to 30 ms. Another problem associated with the application of the classical impact theory

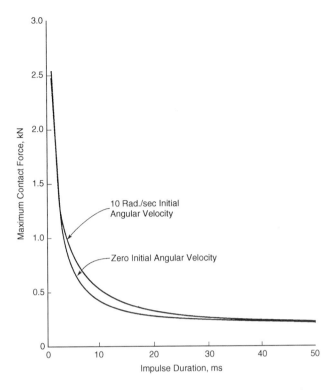

FIGURE 3.12 Maximum contact force vs. impulse duration. (*Source*: Engin, A.E. and Tumer, S.T., *J. Biomechan. Eng.*, ASME, 1993. With permission.)

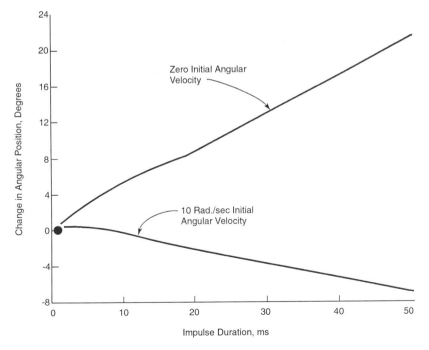

FIGURE 3.13 Change in angular position of the lower leg vs. impulse duration. The result of the approximate solution (classical impact solution) is indicated by (·). (*Source*: Engin, A.E. and Tumer, S.T., *J. Biomechan. Eng.*, ASME, 1993. With permission.)

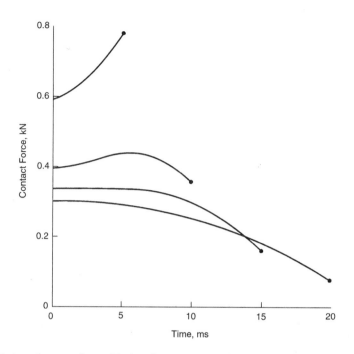

FIGURE 3.14 Variation of contact force with time for various impulse durations (curves are drawn up to the end of durations). (*Source*: Engin, A.E. and Tumer, S.T., *J. Biomechan. Eng.*, ASME, 1993. With permission.)

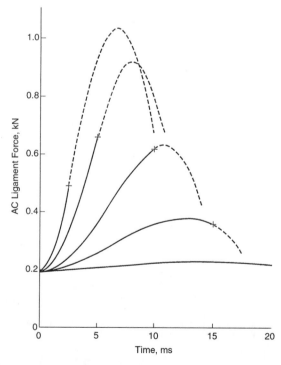

FIGURE 3.15 Variation of anterior cruciate ligament force with time for various impulse durations. The (+) signs indicate the end of impulse duration. (*Source*: Engin, A.E. and Tumer, S.T., *J. Biomechan. Eng.*, ASME, 1993. With permission.)

to the solution of anatomically based models is the difficulty of interpreting the results obtained in the form of impulses. It is shown here that impulse magnitude alone is not sufficient to assess the loading condition at the joint. In fact, such an indication can be quite misleading in that a higher impulse does not necessarily mean higher forces. Finally, the fact that ligament response is not instantaneous entails its exclusion from the classical impact theory, whereas real-time simulations have shown that the ligaments are affected by the impact in comparable magnitudes with contact forces.

3.5 Three-Body Segment Dynamic Model of the Knee Joint

The types of lower-limb activity such as walking, running, climbing, or kicking, and their associated muscle activities are important factors which affect forces transmitted by the knee joint. In most situations, the problem of determining forces, especially their distribution as contact forces between the tibia and femur (tibio-femoral forces), and between the femur and patella (patello-femoral forces) as well as related ligament and muscle forces, is extremely complex.

In an *Applied Mechanics Review* article, Hefzy and Grood[20] discussed both phenomenological and anatomically based models of the knee joint and stated, "To date, all anatomically based models consider only the tibio-femoral joint and neglect the patello-femoral joint, although it is an important part of the knee." In fact, only a few patello-femoral joint models are cited in the literature,[21,37,42] and they are restricted to the study of orientations and static forces related to the patella. Hirokawa's three-dimensional model[21] of the patello-femoral joint has some advanced features over the models of Van Eijden et al.[37] and Yamaguchi and Zajac.[42] Nevertheless, these models consider the patello-femoral articulation in isolation from the dynamics of the tibio-femoral articulation.

In this section, patello-femoral and tibia-femoral contact forces exerted during kicking types of activities are presented by means of a dynamic model of the knee joint which includes tibio-femoral and patello-femoral articulations and the major ligaments of the joint. Major features of the model include two contact surfaces for each articulation, three muscle groups (quadriceps femoris, hamstrings, and gastrocnemius), and the primary ligaments (anterior cruciate, posterior cruciate, medial collateral, lateral collateral, and patellar ligaments). For a quantitative description of the model, as well as its mathematical formulation, three coordinate systems as shown in Fig. 3.16 are introduced. An inertial coordinate system (x, y) is attached to the fixed femur with the x axis directed along the anterior-posterior direction and the y axis coinciding with the femoral longitudinal axis. The moving coordinate system (u, v) is attached to the center of mass of the tibia in a similar fashion. The second moving coordinate system (p, q) is connected to the attachment of the quadriceps tendon, with its p axis directed toward the patella's apex.

Since we are dealing here with an anatomically based model of the knee joint, the femoral and tibial articulating surfaces as well as posterior aspect of the patella and intercondylar groove must be represented realistically. This is achieved by utilizing previously obtained polynomial functions.[11,14,35] For ligaments we use the parabolic and linear force-strain relationship developed by Wismans.[39] The motion of the tibia relative to the femur is described by three variables: the position of its mass center (x_{o2}, y_{o2}), and angular orientation θ_2. Equations of motion of the tibia can be written in terms of these three variables, along with the mass of the lower leg (m), its centroidal moment of inertia (I), the patellar ligament force (F_P), the tibio-femoral contact force (N_2), the hamstrings and gastrocnemius muscle forces (F_H, F_G), the weight of the lower leg (W), and any externally applied force on the lower leg (F_E). The contact conditions at the tibio-femoral articulation and at the patello-femoral articulation are expressed as geometric compatibility and colinearity of the normals of the contact surfaces. The force coupling between the tibia and patella is accomplished by the patella ligament force F_P.

The model has three nonlinear differential equations of motion and eight nonlinear algebraic equations of constraint. The major task in the solution algorithm involves solution of the three nonlinear differential equations of the tibia motion along with three coupled nonlinear algebraic equations of constraint associated with the tibio-femoral articulation. This is accomplished by following solution techniques developed by the author and his colleague which are also described in Section 3.3.[16,17]

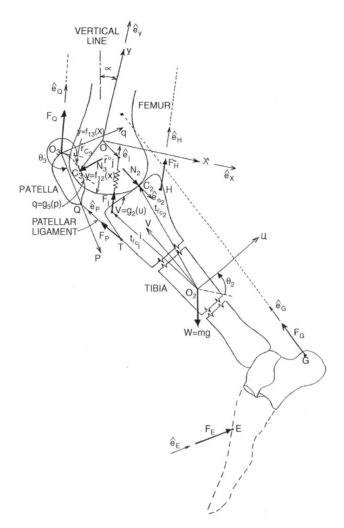

FIGURE 3.16 Three-body segment model of the knee joint showing various coordinate systems, forces, and articulating surface functions.

The kicking type of lower limb activity is a rather complex activity that involves most of the muscles of the lower limb.[25] The model shown in Fig. 3.16 is general enough to include major muscles associated with knee motion, namely, the quadriceps femoris, hamstrings, and gastrocnemius muscle groups.[35] Any dynamic activity can be simulated with this model provided magnitudes, durations, and relative timings of these muscle groups are supplied. In this section, we will present preliminary results for the extension phase of the knee under the activation of the quadriceps femoris muscle group. The force activation of the quadriceps muscle group during the final extension of the knee is taken in the form of an exponentially decaying sinusoidal pulse. For a quadriceps force of 0.1 sec and 2650 N amplitude, the maximum angular acceleration of the lower leg reaches to 360 rad/s^2 which is a typical value for a vigorous lower limb activity such as kicking.

Fig. 3.17 shows for the final phase of knee extension the variations of the tibio-femoral and patello-femoral contact forces with time. The aforementioned quadriceps pulse is applied when the flexion is at 55°. The values in parentheses indicate the flexion angles at the corresponding times; thus, behaviors of the patello-femoral and tibio-femoral contact forces are shown from the flexion angle of 55 to 5.5°. It is quite interesting to note that under such dynamic conditions, the patello-femoral contact force is higher than the tibio-femoral contact force.

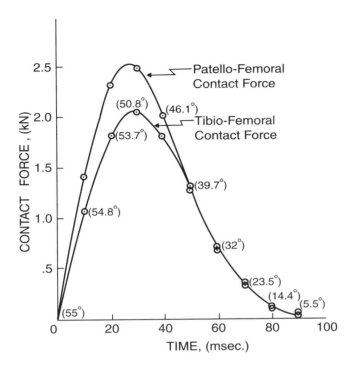

FIGURE 3.17 Variations of tibio-femoral and patello-femoral contact forces with time and flexion angles. (*Source*: Tumer, S.T. and Engin, A.E., *J. Biomechan. Eng.*, ASME, 1993. With permission.

In this section, behavior of both tibio-femoral and patello-femoral contact forces during a kicking type of dynamic activity is presented by means of a two-dimensional, three-body segment dynamic model of the human knee joint. It is a well-established fact that in a class of activities such as stair climbing, rising from a seated position, or similar activities, large patello-femoral contact forces naturally accompany large knee-flexion angles. For these large knee-flexion angles, a rough estimate of the patello-femoral contact force can be easily obtained by considering a simple static equilibrium of the patella with patella tendon force and quadriceps femoris force. According to the static analysis, at full extension of the knee this force is practically zero; as the knee flexes during the above-mentioned activities, the patella-femoral force increases to very high values, e.g., at 135° knee flexion it can reach four to five times body weight. Results presented here indicate that the patella can be subjected to very large patello-femoral contact forces during a strenuous lower limb activity such as kicking even under conditions of small knee-flexion angles. Finally, under such dynamic conditions the patello-femoral contact force can be higher than the tibio-femoral contact force.

References

1. Barnett, C.H., Davies, D.V., and MacConaill, M.A., *Joints: Their Structure and Mechanics*, Charles C Thomas, Springfield, IL, 1949, 111.
2. Bathe, K.J. and Wilson, E.L., *Numerical Methods in Finite Element Analysis*, Prentice-Hall, Englewood Cliffs, 1976.
3. Berne, N., Forces transmitted by the finger and thumb joints in selected hand functions, *Proc. First Meeting European Soc. Biomechan.*, Burny, F., Ed., Acta Orthopaedica Belgica, 1978, 157.
4. Berme, N., Paul, J.P., and Purves, W.K., A biomechanical analysis of the metacarpophalangeal joint, *J. Biomechanics*, 10, 409, 1977.
5. Chao, E.Y., Opgrande, J.D., and Axmear, F.E., Three-dimensional force analysis of finger joints in selected isometric hand functions, *J. Biomechanics*, 9, 387, 1976.

6. Dempster, S.T., The anthropometry of body motion, *Ann. N.Y. Acad. Sci.*, 63, 559, 1955.

7. Engin, A.E. and Moeinzadeh, M.H., Dynamic modeling of human articulating joints, *Proc. Third Internal Conf. Math. Modeling*, 1981, 58.

8. Engin, A.E., Mechanics of the knee joint: guidelines for osteotomy in osteoarthritis, in *Orthopaedic Mechanics: Procedures and Devices*, Ghista, D.N. and Roaf, R., Eds., Academic Press, London, 1978, 55.

9. Engin, A.E., Measurement of resistive torques in major human joints, Report AMRL-TR-79-4, Aerospace Medical Research Laboratory, Wright-Patterson Air Force Base, OH, 1979.

10. Engin, A.E., On the biomechanics of major articulating human joints, in *Progress in Biomechanics*, Akkas, N., Ed., Sijthoff Noordhoff International, Amsterdam, 1979, 158.

11. Engin, A.E. and Moeinzadeh, M.H., Modeling of human joint structures, Report AFAMRL-TR-81-117, Wright-Patterson Air Force Base, OH, 1982.

12. Engin, A.E. and Korde, M.S., Biomechanics of normal and abnormal knee joints, *J. Biomechanics*, 7, 325, 1974.

13. Engin, A.E. and Moeinzadeh, M.H., Two-dimensional dynamic modeling of human joints, *Dev. Theor. Appl. Mech.*, 11, 287, 1982.

14. Engin, A.E. and Moeinzadeh, M.H., Dynamic modeling of human articulating joints, *Mathematical Modelling*, 4, 117, 1983.

15. Engin, A.E. and Peindl, R.D., Two devices developed for kinematic and force data collection in biomechanics: application to human shoulder complex, *Proc. Southeastern Conf. Theor. Appl. Mech.*, 10, 33, 1980.

16. Engin, A.E. and Tümer, S.T., Improved dynamic model of the human knee joint and its response to impact loading on the lower leg, *J. Biomechanical Eng.*, 115, 137, 1993.

16a. Engin, A.E. and Tümer, S.T., Three-body segment dynamic model of the human knee, *J. Biomechnical Eng.*, 1993.

17. Engin, A.E. and Tümer, S.T., An innovative approach to the solution of highly nonlinear dynamics problems associated with joint biomechanics, *ASME Biomechanics Symp.*, 120, 225, 1991.

18. Engin, A.E. and Akkas, N., Application of a fluid-filled spherical sandwich shell as a biodynamic head injury model for primates, *Aviation Space Environmental Med.*, 49, 120, 1978.

19. Frankel, V.H. and Nordin, M., *Basic Biomechanics of the Skeletal System*, Lea & Febiger, Philadelphia, 1980.

20. Hefzy, M.D. and Groos, E.S., Review of knee models, *Appl. Mech. Rev.* 41, 1, 1988.

21. Hirokawa, S., Three-dimensional mathematical model analysis of the patellofemoral joint, *J. Biomechanics*, 24, 659, 1991.

22. Huang, T.C., Roberts, E.M., and Youm, Y., Biomechanics of kicking, in *Human Body Dynamics: Impact, Occupational, and Athletic Aspects*, Ghista, D.N., Ed., Oxford University Press, Oxford, 1982, 409.

23. Kao, R.A., Comparison of Newton-Raphson methods and incremental procedures for geometrically nonlinear analysis, *Comput. Struct.*, 4, 1974, 1091.

24. Kennedy, J.C., Hawkins, R.J., Willis, R.B., and Danylchuk, K.D., Tension studies of human knee joint ligaments, *J. Bone Jt. Surg.*, 58A, 350, 1976.

25. Lindbeck, L., Impulse and moment of impulse in the leg joints by impact from kicking, *J. Biomechanical Eng.*, 105, 108, 1983.

26. MacConaill, M.A., Joint movement, *Physiotherapy*, 50, 359, 1964.

27. Moeinzadeh, M.H., Engin, A.E., and Akkas, N., Two-dimensional dynamic modelling of human knee joint, *J. Biomechanics*, 16, 253, 1983.

28. Nisell, R., Mechanics of the knee: a study of joint and muscle load with clinical applications, *Acta Orthopaed. Scand.*, 56S, 1, 1985.

29. Paul B., Analytical dynamics of mechanics: a computer oriented overview, *Mechanism and Machine Theory*, 10, 481, 1975.

30. Paul, J.P., Forces transmitted by joints in the human body, *Inst. Mechanical Engineers Proc.*, 181, 8, 1967.

31. Radin, E.L. and Paul, I.L., A consolidated concept of joint lubrication, *J. Bone Jt. Surg.*, 54A, 607, 1972.

32. Seirek, A. and Arvikar, R.J., A mathematical model for evaluation of forces in lower extremities of the musculo-skeletal system, *J. Biomechanics*, 6, 313, 1973.

33. Seirek, A. and Arvikar, R.J., The prediction of muscular load sharing and joint forces in the lower extremities during walking, *J. Biomechanics*, 8, 89, 1975.

34. Steindler, A., *Kinesiology of the Human Body*, Charles C Thomas, Springfield, IL, 1964, 62.

35. Tümer, S.T. and Engin, A.E., Three-body segment dynamic model of the human knee, *J. Biomechanical Eng.*, 115, 350, 1993,

36. Trent, P.S., Walker, P.S., and Wolf, B., Ligament length patterns, strength and rotational axes of the knee joint, *Clin. Orthoped. Rel. Res.*, 117, 263, 1976.

37. Van Eijden, T.M., Kouwenhoven, E., Verburg, J., and Weijs, W.A., A mathematical model of the patellofemoral joint, *J. Biomechanics*, 19, 219, 1986.

38. Wahrenberg, H., Lindbeck, L., and Ekholm, J., Dynamic load in the human knee joint during voluntary active impact to the lower leg, *Scand. J. Rehabil. Med.*, 10, 93, 1978.

39. Wismans, J., A., A three-dimensional mathematical model of the human knee joint, doctoral dissertation, Eindhoven University of Technology, The Netherlands, 1980.

40. Wismans, J., Veldpaus, F., Janssen, J., Huson, A., and Struben, P., A three-dimensional mathematical model of the knee joint, *J. Biomechanics*, 13m 677m 1980.

41. Wongchaisuwat, C., Hemami, H., and Buchner, H.J., Control of sliding and rolling at natural joints, *J. Biomechanical Eng.*, 106, 368, 1984.

42. Yamaguchi, G.T. and Zajack F.E., A planar model of the knee joint to characterize the knee extensor mechanism, *J. Biomechanics*, 22, 1, 1989.

4

Techniques and Applications of Scanning Acoustic Microscopy in Bone Remodeling Studies

Mark C. Zimmerman
New Jersey Medical School

Robert D. Harten, Jr.
New Jersey Medical School

Sheu-Jane Shieh
Wayne State University

Alain Meunier
L' Hopital St. Louis

J. Lawrence Katz
Case Western Reserve University

4.1 Acoustic Microscopy Theory, Techniques, and Specimen Preparation

The Physical Interaction between Ultrasound and Solid Media

Acoustic microscopy is akin to optical microscopy — both are techniques that create magnified images of objects. However, where optical microscopy uses waves of light to form its image, acoustic microscopy uses waves of high frequency sound, or ultrasound. The examination of an object with ultrasound creates an image of the object's material and mechanical properties, rather than an image of optical features. The physical interaction of ultrasound with matter is the basis of all acoustic microscopy images.

Presently, almost all acoustic microscopy uses the principle of reflection. When a sound wave is directed onto a large interface, it will be partially transmitted across the interface and partially reflected back toward its source. The reflected portion of the wave is used to determine the mechanical properties of the materials of the interface. The percentage of the wave reflected at the interface depends on the wave's angle of incidence and the acoustic impedance differences of the materials that make up the interface. Snell's law states that for a wave undergoing specular reflection, the angle of incidence equals the angle of reflection. In an acoustic microscope, the source of sound is a piezoelectric transducer which acts as a sender and receiver. Therefore, for the transducer to receive the maximum reflected signal, the transducer is oriented so that the incident wave strikes the interface perpendicularly. The path of the reflected wave thus will also be perpendicular to the interface.

Acoustic impedance is the product of the density and the velocity of ultrasound in the material. The symbol for acoustic impedance is Z and the units are Rayleighs, abbreviated as Rayls. Acoustic impedances are most often reported in MRayls, where 1 MRayl = 10^6 kg m^{-2} s^{-1}. The velocity of ultrasound in biological tissues varies negligibly over a wide range of frequencies; that is, it is nondispersive. The acoustic impedance of a tissue is constant with changing frequency. For reflection at an interface, the pressure of the reflected wave is determined by the difference in acoustic impedance between the two media at the interface. The larger the difference in acoustic impedance at the interface, the greater the pressure of the wave reflected back toward the transducer. Conversely, if the acoustic impedance of the media on each side of the interface is the same, all of the incident beam is transmitted and none will be reflected.

For optimum imaging in acoustic microscopy, the sound pressure after reflection must be maximized, and for this a liquid-solid interface is used. That is, to examine a solid like bone, a liquid couplant is used to propagate sound from the transducer to the bone specimen. To illustrate this, consider sound traveling through water, which has an acoustic impedance of 1.5 MRayls at 20°C, becoming incident on bone, which has an acoustic impedance of approximately 7.8 MRayls. The coefficient of reflection, R, is computed by the formula:

$$R = \frac{Z_2 - Z_1}{Z_2 + Z_1}$$

where Z_1 and Z_2 are the acoustic impedances of the two interfacial media, water and bone. Expressed as a percentage, the reflected wave will have 68% of the sound pressure of the incident wave. However, this large amplitude is not the amplitude of the wave received by the transducer. As the reflected wave travels through the water on the return path toward the transducer, it is attenuated and suffers a loss in intensity. Moreover, this attenuation increases with increasing frequency. This is a practical limitation of reflection acoustic microscopy which prevents it from being used to study softer materials such as soft biological tissues which have acoustic impedances only slightly higher than water.

Attenuation in the second media at the interface is not a concern in this case. Although the transmitted wave traveling through the second media will undergo attenuation, the reflected wave will not be affected. It is evident upon inspection of the reflection coefficient formula that the acoustic impedance of the second material at an interface may be determined if the acoustic impedance of the first material is known and the reflection coefficient measured.

The Relationship between Acoustic Impedance and Mechanical Properties

Once the acoustic impedance of a material is determined, the relation to the mechanical properties of a material may be assessed. As stated earlier, acoustic impedance is a function of material density and velocity. It will now be shown that the mechanical properties of stiffness, in particular, the bulk modulus, Young's modulus, and the shear modulus are also functions of the same material density and velocity.

When the wave equation which describes the displacement of a harmonic wave is substituted into the general equation of motion of a continuum, a relation is established between the velocity of propagation and stress since, in the equation of motion, the force exerted on a solid is the gradient of the stress. (For a full derivation, the reader is referred to Beyer and Lechter and Kolsky.[1,2]) If Hooke's law for a linear elastic solid is also substituted into this equation, the equation:[*]

$$C_i = \sqrt{k + \tfrac{4}{3} G / \rho}$$

results for isotropic, homogeneous media. This is the equation for the bulk longitudinal dilatational velocity C, in an infinitely extended medium of bulk modulus k, shear modulus G, and density ρ. In a bar subject to a longitudinal wave, if the wavelength is long compared to the lateral dimensions, the entire cross section is subject to uniform stress and displacement and the velocity of propagation extensional longitudinal is the bar wave velocity C, where:[**]

$$C_L = \sqrt{E/\rho}$$

and E is Young's modulus. The distinction in the two velocity equations exists because of the difference in the relative dimensions of the wavelength and the media. In the first case, the propagation of longitudinal waves subjects the infinitely extended media to compression and shear, and thus the shear modulus and the bulk modulus are involved.

These two velocity equations and the equation of acoustic impedance show that the material moduli and the acoustic parameter Z are dependent on the same physical constants, namely, density and velocity. Therefore, a direct proportionality exists between the acoustic impedance and the mechanical properties of a material. With acoustic microscopy, a measure of the mechanical properties of a material may be determined by invoking their proportionality with acoustic impedance. Moreover, acoustic microscopy affords a great advantage over traditional mechanical testing — the ability to determine mechanical properties on a submillimeter level. This ability is derived from the use of ultrasound that can be focused to spot sizes smaller than one millimeter.

The Spatial Resolution of the Acoustic Microscope

The fundamental components of the acoustic microscope are the transducer and lens. A single flat element piezoelectric transducer acts as sender and receiver, generating and acquiring pulses of waves. By refraction, a spherical lens focuses the plane waves that are excited by the transducer. The intensity of the ultrasound beam reaches its peak at a distance from the lens equal to its focal length. In the focal region, the beam narrows, and the energy is concentrated into a smaller area. The size of this area is the spatial resolution of the beam. The transducer is positioned above the media to be interrogated at the focal length to achieve maximum reflection of the incident wave.

Transducers used in acoustic microscopy differ mainly in their frequency, focal length, size, and the ratio of the focal length to aperture (known as the F number). A transducer with a focal length of 32 mm and diameter of 16 mm has an F number of F2. The acceptance angle of a transducer is the maximum angle relative to its long axis at which acoustic waves are sent or received. The sine of the acceptance angle θ_{max} is one-half the radius of curvature divided by the diameter or

$$\sin \theta_{max} = \frac{1}{2} F$$

Particular transducers are designed with an acceptance angle below the critical cutoff angle for which Rayleigh rays are generated to ensure that only specular reflection exists.

The spatial resolution of a transducer depends on the operating frequency, coupling medium, and transducer properties. The relation between resolution and frequency is demonstrated in the equation:

[*]Uniaxial with lateral constraints.

[**]Equivalent to uniaxial mechanical test.

$$\lambda = c/f$$

where λ is wavelength, c is velocity, and f is frequency. The smaller the wavelength (the higher the frequency), the finer the spatial resolution will be. This equation assumes that the ultrasonic wave is comprised of a single frequency.

In reality it is impossible for a pulsed transducer to generate a single frequency signal because of inertial effects. The bandwidth of a signal is the range of frequencies in the ultrasound pulse — shorter pulses will have larger bandwidths. In acoustic microscopy, short pulses are necessary to ensure that incident and reflected waves do not interfere. A transducer that has short pulse lengths has a short "ring down time" because it oscillates through fewer numbers of cycles per excitation.

A Historical Perspective of the Study of Bone with Acoustic Microscopy

The first practical scanning acoustic microscope was developed and built for materials analysis by Lemons and Quate.[3] Jipson and Quate[4] improved the acoustic microscope, enabling it to obtain resolutions comparable to those of an optical microscope. Since then, scanning acoustic microscopy has become a major nondestructive evaluation tool involving imaging and the characterization and detection of defects in structural materials.[5,6] As the acoustic waves travel through the specimen, it can be focused at different levels within the sample, permitting the technique to produce images which feature the surface or the subsurface of a specimen.

Although acoustic microscopy has been used for over 3 decades in industry to analyze solid materials, only recently have scientists applied scanning acoustic microscopy to the fields of biology and medicine, where "soft" materials are prevalent. Experiments involving the technique have been used to study the acoustic velocity, acoustic attenuation, density, and thickness of tissues, and even cells.[7-12] Because of their structures and calcified natures, bones and teeth have higher acoustic reflectivities than soft tissue, and to date most studies have used acoustic microscopy for the qualitative examination of bone.[10,11,13-17]

Some very interesting high resolution qualitative observations have been made by Katz and Meunier.[18] Their findings will be discussed in more detail in Section 4.2.

To measure the acoustic impedance of a material, one collects the amplitude of the first echo of a longitudinal sound wave reflected from the surface of the specimen. Meunier et al.[14] have verified this technique with bone as well as many other materials. By varying transducer design, other sonic waves can be generated and used for material properties analysis. Gardner et al.[19] have demonstrated that with a technique of defocusing Rayleigh waves, the material properties of many materials can be measured. Rayleigh waves are surface waves; however, bone poorly supports Rayleigh wave propagation. High signal attenuation occurs because of the porosity of bone. Low Rayleigh velocities result, and given the angle at which the signal returns to the transducer, it is not possible for the transducer to receive the signal. The same group measured surface skimming waves from horse radial bone. A high frequency point focus lens was used and surface waves were measured every 150 microns in a line. The averaged velocities were used to calculate a mean elastic coefficient.

Other investigators have also used the velocities measured with acoustic microscopy to calculate the material properties of bone. Hasegawa et al.[20] used an acoustic microscope to measure the velocity of biopsied human bone specimens. They observed a significant decrease in the velocity of sound in osteoporotic bone compared to age-matched normal bone. Three other recently published studies have addressed the material property contributions that mineral and collagen provide for bone, and all three research groups utilized a scanning acoustic microscope. Broz et al.[21] used a number of material analysis techniques (including acoustic microscopy) to analyze the material properties of serially demineralized bone. All the techniques demonstrated decreased material properties with demineralization. Turner et al.[22] also used an acoustic microscope to study the collagen and mineral anisotropy of bone. They also compared the elastic constants generated with bulk transmission ultrasound against the constants generated with velocities measured with an acoustic microscope. Orthotropic symmetry was assumed, and a good agreement with both techniques was observed, except for specimens oriented 30° to the longitudinal axis of the bone.

This deviation supports the hypothesis that the principal orientation of the secondary lamellar bone collagen is 30° to the long axis of the bone. This concept was further supported by the anisotropy ratio data generated in this study. The most recent work from Turner's laboratory[23] makes use of the collagen and mineral orientation data in a composite model for the ultrastructure of osteonal bone. A two-phase composite model was developed and the acoustic data played a key role in testing hypotheses regarding amounts of intrafibular vs. extrafibular mineral and collagen orientation.

A scanning acoustic microscope with a similar transducer was used in the studies by Hasegawa,[20] Turner,[22] Pidaparti,[23] and Broz.[21] Hasegawa, Turner, and Pidaparti reported a resolution of 60 microns and a precision of 0.3 to 0.5%. Broz et al. reported that their transducer produced a 100 μm resolution and scans were done with 50 micron stepping. In all four studies, acoustic velocity measurements were made at multiple locations on each specimen and then averaged for a mean velocity. Bulk density measurements were made using Archimedes' principle for the entire specimen and bulk elastic stiffness coefficients were calculated. These data were used to calculate the bulk elastic stiffness coefficients of the specimens. Thus, a relatively high resolution velocity measurement tool was used to measure multiple velocities, and a mean velocity was used to calculate bulk properties in all four studies.

Specimen Preparation

In all the experiments to be described, fresh or imbedded bone was tested. The imbedded bone goes through a dehydration and clearing process before imbedding in polymethylmethacrylate. After imbedding, the specimens are sectioned with a diamond saw and polished with fine sized grit until a 600 grit (approximately 15 μm) finish is attained. In all scanning experiments, deionized water was used as a couplant. An optical reflection microscope was used to examine the quality of the polishing and detect any defects or cracks that would affect the acoustic measurements. Once the specimen was aligned in the instrument, the parallelism assured that the wave front was always perpendicular to the specimen. Finally, each section was placed in an ultrasonic cleaner for 5 minutes to clear the bone of any grit that may have entered the pores during the polishing process.

The Acoustic Microscope

The acoustic microscope used in the low frequency studies consisted of a Panametrics 5052UA ultrasonic analyzer, Matec SR-9000 pulser receiver, a differential amplifier, and a 12-bit A/D converter. The specimens were mounted in a water bath and scanned in a raster fashion with a 50 MHz spherically focused transducer (Panametrics Inc., V3204, 12.7 mm focal length). A microprocessor served multiple functions. It drove the scanning of the specimen, it acquired the data through an A/D converter (a 256×256 array of points per image is stored), and it presented the results visually.

The resolution of the images ranged from 20 to 140 microns depending on the stepping distance. The 50 MHz transducer used in these studies had a spot size of 20 microns, which dictated the maximum resolution of this system. Software developed for this application produces a pseudo-color map of the acoustic impedance of the surface of the specimen in two or three dimensions, where either color (2D) or profile (3D) is indicative of reflected acoustic impedance. The depth of penetration of the analyses varies from roughly 20 to 80 μm in well-mineralized bone. The depth of penetration is approximately equal to one wavelength which is equal to the speed of sound in water divided by the frequency of the transducer (1500m/s divided by 50MHz). Thus, in materials with higher acoustic velocities, e.g., more mature bone, the depth of penetration is greater.

This program also allows quantitative mapping of the coefficient of reflection, area, and thresholding with numerical data in an easily accessible data array. The acoustic impedance is determined from the acquired amplitude information through a calibration scheme using reference materials. Two homogeneous materials of known acoustic impedance are routinely scanned under the same operating conditions as the experimental specimens. Typically, glass with an acoustic impedance of 13.56 MRayl and plastic (Plexiglas®) with an acoustic impedance of 3.23 MRayl are used. The impedances of the glass and plastic were previously determined in a separate transmission ultrasonic experiment, and verified by comparison with published data.[24]

A very simple experiment was conducted to evaluate the validity of our calibration scheme and the accuracy of our acoustic impedance measurements with our acoustic microscope.[25] This experiment entailed the sectioning and processing of bovine femoral bone. The bone was scanned, and the impedances measured using our calibration algorithm. A standard ultrasonic transmission technique[26] was used to also measure the acoustic impedance of the specimens. Using this technique, the impedance was simply the product of the transmission velocity and the density of the specimen. As shown in Fig. 4.1, an excellent correlation was observed between the two techniques ($r^2 = 0.976$). This verified that the system accurately measured impedance and the calibration scheme was adequate.

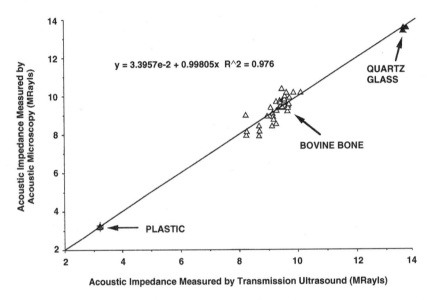

FIGURE 4.1 A graphical representation of acoustic impedance of bovine bone measured with a scanning acoustic microscope vs. a standard ultrasonic transmission technique. An excellent correlation was observed for the two techniques.

A secondary finding in this study was the effect of plastic imbedding on the acoustic properties of bone. The fresh bovine specimens scanned for the calibration experiment were then imbedded in polymethylmethacrylate (PMMA) as previously described, repolished, and scanned again. A small but statistically significant increase was observed for the imbedded specimens (0.4 MRayls). The increase was most likely caused by the filling of the pores of the bone with PMMA. This small increase can be eliminated by simply thresholding the window of acoustic properties to be analyzed. Thus, if the lower limit of a threshold window is set above the impedance of PMMA (3.18 Mrayls), then the effects of imbedding are eliminated.

The high frequency studies were performed using an Olympus UH3 SAM in the burst mode (a train of 10 or so sinusoids) at frequencies from 200 MHz (resolution 15 μm) to 1 GHz (resolution 1 μm). Unfortunately the depth of penetration is reduced considerably in well-mineralized bone at these frequencies compared to the 20 to 80 μm cited above; at 1 GHz it is down to a few μm. Therefore, the high frequency studies were mostly of the surface properties.

4.2. Acoustic Studies of Cortical Bone Material Properties

Low Frequency

The purpose of these experiments was to evaluate the acoustic properties of cortical bone. In the three studies to be described, specimens were taken from the femurs of humans or bovine species. The femur was chosen as the source of this tissue because of its biomechanical function, size, and the fact that its mechanical and material properties have been intensely investigated in previous studies. Additionally, of all the locations throughout the appendicular skeleton, cortical bone from the shaft of the femur is typically the most uniform and dense. Therefore, samples from this region are easier to fabricate, and the determination of averaged or bulk properties involves less experimental error due to the consistent nature of this tissue.

In the first study, six pairs of human femurs (3 males and 3 females, mean age of 85.6) were harvested and frozen. The central portion of each femur, 70% of the biomechanical length as defined by Bloebaum et al.,[27] was divided into ten 10 mm sections which were transversely cut beginning at the distal border of the lesser trochanter. Section 1 corresponded to the section closest to the hip and section 10 to the section closest to the knee. Fig. 4.2 shows the anatomic positions of the sections and the average impedance/section.

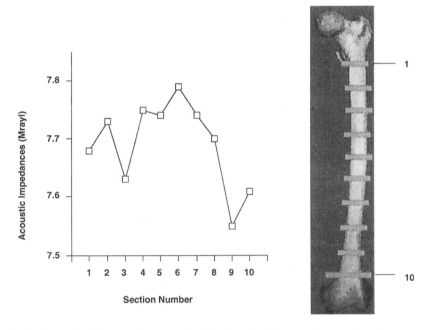

FIGURE 4.2   An image of the human femora and section locations for scanning acoustic microscopy.

Graphic images were stored as 256×256 arrays at a resolution of 140 microns. Acoustic impedance was measured for the entire cross section as well as for four quadrants within each cross section of bone. These four quadrants were designated with respect to anatomic position: anterior, posterior, medial, and lateral. High resolution scanning was also accomplished for a number of specimens consisting of a 1024×1024 array of data points with 20 micron resolution. These images were used primarily for a qualitative analysis of local properties.

Of the four factors analyzed (side, quadrant, sex, and cross-section number), only the cross-section number and the quadrant were significant factors affecting the impedance level ($p = 0.0018$ and $p = 0.0001$, respectively). Longitudinally, the acoustic properties of the cross sections peaked at mid-shaft. For instance, levels for sections 3, 9, and 10 were significantly lower than that for section 6. From a quadrant perspective, a Tukey's multiple comparison test showed the posterior quadrant to have a lower

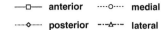

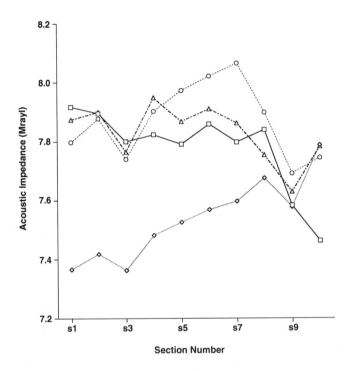

FIGURE 4.3   A graph highlighting the mean acoustic impedances within the human femoral cross sections.

impedance than the other quadrants (Fig. 4.3). Within the transverse sections, the posterior quadrant had statistically significantly lower impedances for the six most proximal sections. Proceeding distally, the impedance values of the four quadrants appear to converge on one another. It is interesting to note that the impedance of the anterior quadrant was significantly less than those of the three other quadrants in the most distal section.

The impedance variations observed at the different levels along the length of the femur, as well as within the cross sections, mirror the longitudinal elastic coefficient (C_{33}) and density variations obtained previously by Ashman et al.[26] and Meunier et al.[28] Ashman only analyzed the middle 40% of the femur, which corresponds to sections 3 through 7 in the present study. They observed that the density and elastic stiffness coefficient were greater at the 50, 60, and 70% levels of the femur relative to the 30 and 40% levels. In the present study, those parameters for levels 4, 5, 6, and 7 were greater than those for the other levels. Ashman also observed a statistically significant decrease in the elastic stiffness coefficient and the density of the posterior quadrant relative to the other quadrants.

Fig. 4.4 is an acoustic scan of a cross section taken from the left femur of an 86-year-old male. This example shows very uniform acoustic properties. The mean acoustic impedance of the most proximal segment (section 1) is 7.50 MRayls and the lateral zone shown by the box has an average impedance of 7.68 MRayls. Local variations in bone mineralization and structure result in local impedance differences which are shown by yellow and orange colors. The higher resolution scan of the local region has an approximate resolution of 20 microns.

Fig. 4.5 is an example of scans taken from bones with varying porotic and impedance properties. The left scan is a severely osteoporotic section taken from a 88-year-old female with an average impedance of 7.78 Mrayls. The image on the right is taken from the femur of an 86-year-old male, and has an average impedance of 7.98 MRayls. There is a striking difference in bone volume and color histograms show the

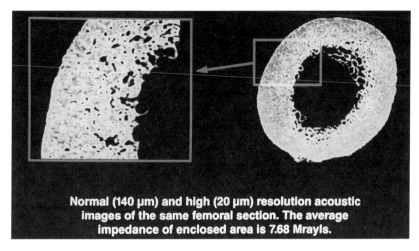

Normal (140 μm) and high (20 μm) resolution acoustic images of the same femoral section. The average impedance of enclosed area is 7.68 Mrayls.

FIGURE 4.4 A low and high resolution acoustic scan of an 86-year-old male. The two resolutions are highlighted and the histogram shows the variation in the acoustic properties.

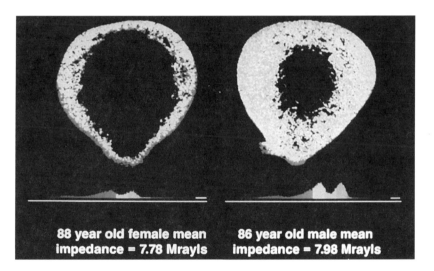

88 year old female mean impedance = 7.78 Mrayls 86 year old male mean impedance = 7.98 Mrayls

FIGURE 4.5 Two transverse cross sections taken from an 88-year-old female and an 86-year-old male at level 5. The female femur was severely osteoporotic while the male femur has good bone quality with an impedance histogram that shows greater volume and impedance.

volume of bone of varying properties. Both scans have bimodal distributions of acoustic properties, although a significant shift to the right, or a general increase, is apparent in the non-porotic bone from the male. Also of note is the lower impedance periosteal bone of the porotic femur, and the amount of lower impedance or "green" bone throughout the section. Additionally, islands of higher impedance bone are seen throughout the section with the antero-lateral section recording some of the highest values. This image demonstrates the highly dynamic nature of osteoporotic bone and the subsequent disproportionate property distribution that is a function of bone remodeling.

This was also observed by Meunier et al.[28] when bone sections that exhibited very low average acoustic characteristics still presented numerous locations with high acoustic properties. They observed that the decrease in the elastic properties due to bone remodeling is not a uniform process throughout the cortex but occurs in preferential areas while some locations maintain their properties for long periods of time. This concurs with previous results reporting that osteoporotic tissue is quantitatively deficient (low in volume), but remains qualitatively normal.[29]

Fig. 4.6 is another example of two different femoral sections taken from the same cadaver. These images demonstrate two things: the dynamic differences that can be observed from intracadaveric comparisons and the results of a circular averaging technique. A technique was developed by which the circular cross section was divided into 36 radial sections by calculating the properties of the bone at every 10°.[28] An algorithm was then used to calculate a weighted pixel acoustic impedance based on the acoustic properties of all the pixels that surround an individual pixel. The distance between the periosteum to the endosteum was divided by three and the pixels that fell within the three locations were averaged to generate periosteal, middle, and endosteal impedances. A total average was also calculated. Again, significant differences in properties were observed in the endosteal and periosteal locations of the osteoporotic specimen. These results were accentuated by the graphs of the acoustic properties presented on the right. The values on the y axis represent the acoustic impedance on a 0 to 256 scale.

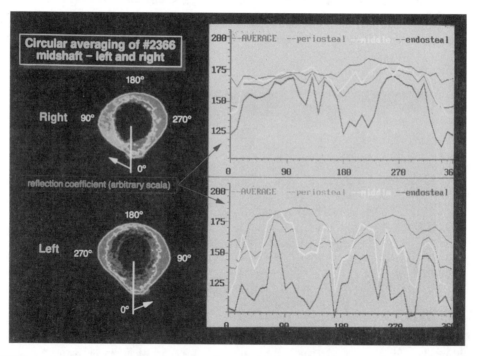

FIGURE 4.6 Two transverse cross sections taken from the same cadaver at the same level. This image demonstrates the variation that can be observed within the same individual. It also highlights the "circular averaging" technique developed by Meunier et al. There is a greater degree of variation observed for bone from different regions within the osteoporotic cross sections. The other section has a less erratic pattern of bone properties.

The right section shows that the periosteal, middle, and average values are relatively constant across the specimen with a decrease in the posterior location of the femur. It is interesting to note that the endosteal properties of this bone have a more irregular pattern and there is a fairly large decrease in the anterior endosteal zone. The porotic femur reveals a much more erratic image, with both the endosteal and middle properties varying widely across the specimen. The average also follows this volatile nature of the middle and endosteal values. Also of note is the general shift of the curves in the direction of reduced properties for the porotic section. This relatively simple technique of identifying the properties within different locations of the section provides a powerful graphic image of the range of remodeled acoustic properties that occur in a diseased state.

In the second study, twenty pairs of human femora were sectioned and scanned.[28] The forty femora were also tested fresh and were sectioned with a diamond saw. Only one section was prepared from the midshaft of the diaphysis of each. The acoustic properties were measured using both a reflection acoustic microscope and a well-published transmission ultrasound technique.[26] Four quadrants from each specimen were evaluated as in the previous study (anterior, posterior, medial, and lateral). The acoustic scanner was slightly different from the one used in the aforementioned system in that a 30 MHz transducer was used. As shown in Fig. 4.7, a second order relationship was established for the reflection coefficients and the elastic stiffness coefficients. An r value of 0.99 demonstrated an excellent agreement between these techniques. This finding coupled with the results shown in Fig. 4.1 presents a strong argument that acoustic impedance is directly related to a material property of bone. The elastic stiffness coefficients are used in the calculation of Young's modulus,[30] and in this case, C_{33} is the most important coefficient in the calculation of Young's modulus.

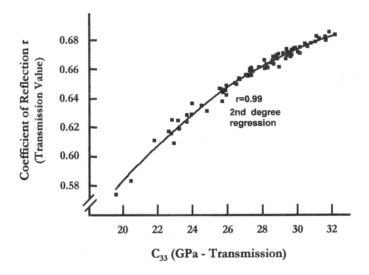

FIGURE 4.7 The relationship between the reflection coefficient and the elastic stiffness coefficient for human femoral bone.[28] An excellent correlation was observed, $r^2 = 0.99$.

Finally, in a most recent study, the relationship between impedance and modulus was tested.[31] Bovine bone was machined and soaked in solutions varying concentrations of NaF. It is hypothesized that the flouride ions replace the hydroxide (OH^-) ions within the hydroxyapatite mineral component of bone. In all, 27 bovine bone specimens were machined from the medial and lateral quadrants of the central 10 mm of each femur. The specimens were formed into flattened dumbbell shapes. The bones were then separated into 50 cc of one of four treatment solutions. Group 1 (n = 6) samples were placed into a solution of 0.145 M NaCl, Group 2 (n = 6) solutions contained 0.145 M NaF, Group 3 (n = 6) solutions contained 0.5 M NaF, and Group 4 (n = 9) solutions consisted of 2.0 M NaF. The samples were set in a water bath shaker for 3 days at 37°C. Fluoride content and density were measured after treatment.

The specimens were mechanically tested to failure in tension. Under constant irrigation with their respective solutions, tests were performed under stroke control at a strain rate of 2.65×10^3 s⁻¹. Deformation was measured using an extensometer with a gage length of 12.5 mm. The specimens were then prepared for SAM analysis. Two scans were made from the gaged length of each specimen in a deionized water bath at room temperature. The images generated had a pixel size of 20 μm. The average impedance of each cross section was measured.

The data demonstrated a good correlation between Young's modulus and acoustic impedance. The mean impedance and modulus for each specimen are plotted in Fig. 4.8. The results were best described

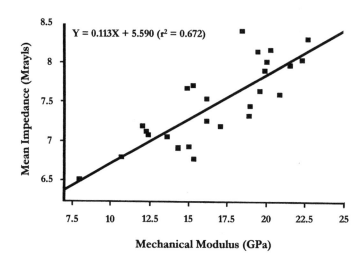

FIGURE 4.8 A graph of the mean acoustic impedance vs. the Young's modulus for bovine bone that has been soaked in solutions of varying concentrations of sodium fluoride.

by a linear relationship with an $r^2 = 0.672$. Increasing concentrations of NaF were seen to result in concomitant decreases in modulus and impedance. The NaF treatments appeared to affect the bone stiffness by diffusing into the samples and altering the bone mineral. This was evident in the SAM images, in which a border of bone with decreased impedance surrounded a core of bone with normal acoustic properties. The width of this border was greater with increasing concentrations of NaF. In fact, with 2.0 M NaF, many samples were completely and uniformly penetrated.

Although the relationship is not quite as strong as was observed for the elastic stiffness coefficient in Fig. 4.7, this experiment further supports the relationship between the acoustic impedance and the mechanical properties of bone. Two factors that may have played a role in this weaker relationship include eight other elastic stiffness coefficients are not taken into account with respect to the impedance in one direction; and only two sections within an entire gage length of bone were tested. Additionally, variations in the bone material within the gaged lengths would have also existed. These variations would arise due to the naturally occurring heterogeneities of the bone itself, as well as local differences in flouride ion penetration.

High Frequency

The initial studies were performed using the Olympus UH3. on canine femoral cortical bone.[18] Specimen preparation was performed as described above. Fig. 4.9a is a SAM micrograph obtained of a cortical area of a canine femur cut transverse to the bone axis using a 400 MHz burst mode lens (resolution 3.5 μm) over a 1 mm lateral scan width. Three important new findings were observed. First, the outermost lamellae of each Haversian system (secondary osteon) had the darkest gray levels. The dark gray level here implies lower density and/or elasticity; light gray level implies higher density and/or elasticity. Second, wherever two or more adjacent osteons were abutting, the gray levels of their outermost lamellae appeared to interdigitate. Finally, within an osteon all the lamellae did not have the same gray levels.

Indeed, Fig. 4.10a is a higher frequency scan of the osteon seen in the lower left-hand portion of Fig. 4.9a, and the gray levels of the outermost lamellae are the darkest, while the interior lamellae appear to vary between light and dark, implying variations in density and/or elasticity in adjacent lamellae (600 MHz burst lens, resolution 1.7 μm, lateral scan width 250 μm). Similarly, Fig. 4.11a is a triangular-like intersection of the three osteons indicated by the junction of the arrows seen on Fig. 4.9a (600 MHz burst lens, resolution 1.7 μm, lateral scan width 100 μm). The interdigitation overlap of the darkest gray levels is clearly seen, implying overlap of density and/or elastic properties.

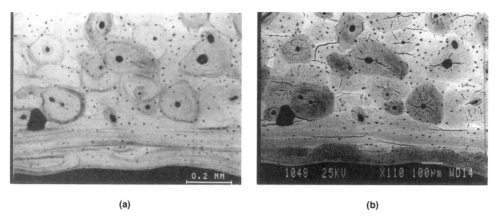

(a)   (b)

FIGURE 4.9a   400 MHz (aperture angle 60°, burst mode, full scale x dimension is 1 mm) SAM micrograph of a portion of canine femoral cortical bone cut transverse to the bone axis. Of particular interest is the lower reflectivity (darker shade of gray) observed for the outermost lamellae of each Haversian system.

FIGURE 4.9b   Backscattered electron SEM micrograph of the same region shown in Fig. 4.9a (cracks are drying artifacts). The Haversian systems are well defined and separated from neighboring Haversian systems.

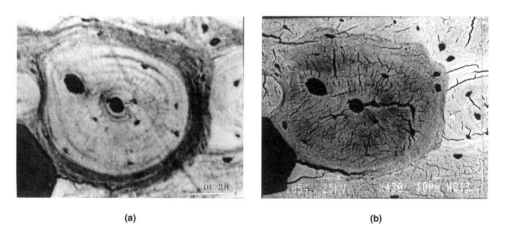

(a)   (b)

FIGURE 4.10a   600 MHz (aperture angle 120°, burst mode, full scale x dimension is 250 μm) SAM micrograph of a large Haversian system in lower left quadrant of Fig. 4.9a clearly showing the lower reflectivity of the outermost lamellae and the apparent interdigitation of gray levels in the upper right corner. The second lumen slightly above and to the left of the Haversian canal appear to be a second canal based on the appearance of lamellae surrounding the opening.

FIGURE 4.10b   Backscattered electron SEM micrograph of the same region shown in Fig. 4.10a. The Haversian system is well defined and separated from its neighboring Haversian systems.

To determine whether these outermost lamellae do indeed belong to distinctly different osteons, a second technique delineating structural features must be used. In this case, backscatter scanning electron microscopy (BSEM) was chosen. In the rush to make the structural observations, the proper drying procedures to limit cracking artifacts were not used. Fortunately, the crack artifacts do not obscure the structural information as seen on Figs. 4.9b, 4.10b, and 4.11b, the BSEM micrographs of the same areas seen on the SAM micrographs, Figs. 4.9a, 4.10a, and 4.11a, respectively. It is clearly seen on the SEM micrographs that the osteons are structurally distinct but their outermost lamellae gray levels do indeed interdigitate.

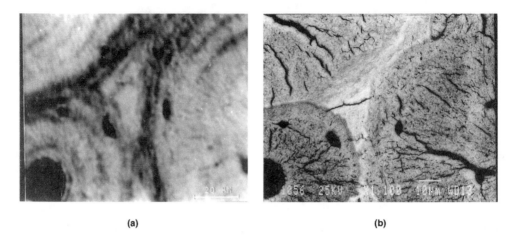

(a)   (b)

FIGURE 4.11a   600 MHz (aperture angle 120°, burst mode, full scale x dimension is 100 mm) SAM micrograph of the selected area between three Haversian systems located at the intersection of the arrows in the lower right quadrant of Fig. 4.9a. The reflectivities of the outermost lamellae of each Haversian system are seen to be comparable, creating what appears to be an interdigitation of these lamellae even though the backscatter electron SEM (Fig. 4.1b) clearly indicates each Haversian system to be distinct and separated from its neighbors.

FIGURE 4.11b   Backscattered electron SEM micrograph of the same region shown in Fig. 4.11a. The Haversian systems are well defined and separated.

These observations on canine bone indicated that it was imperative to perform a similar study on fresh human femoral cortical bone.[18] Fig. 4.12a is a SAM micrograph of a specimen of human femoral cortical bone described by Katz and Meunier[18] (600 MHz burst mode lens, resolution 1.7 μm, lateral scan width 1 mm). Fig. 4.12b is the B SEM micrograph of the same area seen on Fig. 4.12a. Proper drying techniques were used so that crack artifacts were held to a minimum. The same three critical observations made in the canine bone study were made for human bone as well. Indeed the variations in gray levels in alternate lamellae can be observed more dramatically in Fig. 4.13a which is a dark elliptically shaped osteon in the upper left-hand quadrant of Fig. 4.12a (600 MHz burst mode lens, resolution 1.7 μm, lateral scan width 250 μm).

In order to ensure that the specimen surface is both free of gross irregularities and is aligned properly with respect to the lens, a special type of SAM image is run, a so-called x-z image. Here, the lens is defocused a fixed distance, z, below the surface and is then moved upward in the z direction while vibrating along a line in the x direction. This provides an acoustic interference profile in the x-z plane that is due to the interaction of the surface and longitudinal acoustic waves providing acoustic information that can be related to the specimen's elastic properties along the scanned line.

Fig. 4.13b is an x-z SAM micrograph with the same 600 MHz lens taken along the line shown on Fig. 4.13a. The small variations in width of the band reflect the variations in elastic properties among the alternating lamellae as well as possible small height artifacts caused during the polishing process. A calibration system has been developed using a number of materials whose impedance values vary over a wide enough range of values to allow for obtaining elastic moduli for materials varying from low impedances as in soft tissues and polymers to high impedances as in metals and ceramics.

The gray level variations of alternating lamellae observed in cortical bone were also seen in trabecular bone. Fig. 4.14a is a SAM micrograph of a portion of cancellous bone from a sheep femoral condyle (600 MHz burst mode lens, resolution 1.7 μm, lateral scan width 250 μm). Fig. 4.14b is a SAM micrograph taken with the same lens over a lateral scan width of 100 μm showing the region surrounding the lunate-shaped defect seen in the lower left-hand corner on Fig. 4.14a. Not only are the variations in gray level observed, but the effect of the defect on lamellar organization and properties is also clearly delineated.

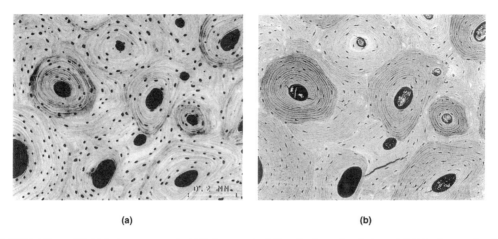

(a)   (b)

FIGURE 4.12a   600 MHz (aperture angle 120°, bust mode, full scale dimension is 1 mm) SAM micrograph of a portion of human femoral cortical bone cut transverse to the bone axis. The same reduced gray levels of the outermost lamellae and their apparent interdigitation as seen for the canine bone (Figs. 4.9a, 4.10a, 4.11a) are observed here also.

FIGURE 4.12b   Backscattered electron SEM micrograph of the same region shown in Fig. 4.12a. The same clear delineation of individual Haversian structures observed for canine bone (Figs. 4.9b, 4.10b, 4.11b) is observed here also.

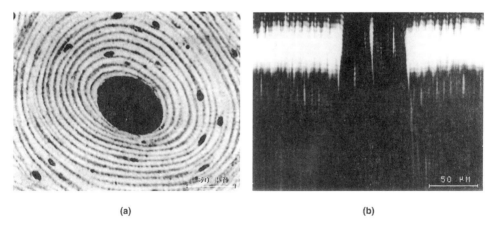

(a)   (b)

FIGURE 4.13a   600 MHz (aperture angle 120°, burst mode, full scale x dimension is 250 μm) SAM micrograph of the dark Haversian system seen in the upper left-hand quadrant of Fig. 4.12a clearly showing the variations in gray levels of alternate lamellae.

FIGURE 4.13b   x-z SAM image along the line through the mid-point on Fig. 4.13a (full scale x is 250 μm). The level white band shows the sample was flat and well aligned perpendicular to the lens. Slight incursions in the flat surface reflect, in part, polishing artifacts.

The use of high resolution SAM studies on soft tissues is exemplified in Fig. 4.15, a 400 MHz burst mode micrograph of a sheep meniscus taken over a 1 mm lateral scan width. The collagen fibrillar organization is clearly depicted.

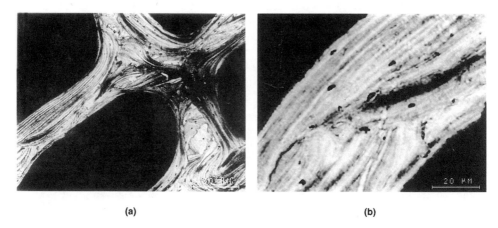

(a)   (b)

FIGURE 4.14a   600 MHz (aperture angle 120°, burst mode, full scale x dimension is 250 μm) SAM micrograph of a trabecular portion of a specimen of sheep femoral condyle showing clearly the same alternations in gray levels in the lamellar structure as observed in the osteonic lamellae.

FIGURE 4.14b   600 MHz (aperture angle 120°, burst mode, full scale x dimension 100 μm) SAM of the lower left-hand quadrant of Fig. 4.14a. The disruption of the lamellar structure and properties at the tip of the lunate-shaped defect is observed.

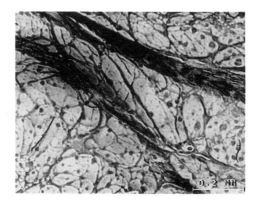

FIGURE 4.15   400 MHz (aperture angle 120°, bust mode, full scale x dimension is 1 mm) SAM micrograph of sheep meniscus showing the collagen fibrillar organization.

4.3   Acoustic Microscopy and Biomaterials Analysis

Acoustic Properties of Remodeled Bone about Metal and Ceramic Coated Metallic Prostheses

Uncemented porous-coated titanium hip implant systems are presently available for total joint arthro-plasty. Macrotextured titanium hip implant surfaces have been investigated and include sintered com-mercially pure titanium (CPTi) beads, diffusion bonded titanium fiber metal pads, and plasma-sprayed CPTi coatings. The purpose of these porous-coated or roughened implant surfaces is to achieve an interlock with the surrounding bone, i.e., the surrounding bone will infiltrate the recesses of the surface macrotexture and provide axial and torsional stability.

To manufacture sintered CPTi beads, layers of spherical CPTi beads are positioned on the femoral component with a binding substance. The femoral component with the applied binding substance–bead mixture is then subjected to the sintering process. Sintering is a high temperature process that dissipates the binding substance and fuses the beads to each other and to the femoral component. The high temperatures that the femoral component substrate is exposed to can significantly decrease the fatigue strength of the implant system.[32] The strength of the bond of the porous coating to the femoral component substrate is controlled by temperature and length of heating. The porosity is controlled by the sizes of the spherical beads.

Fiber metal pads are produced from wire that has been cut, kinked, and formed in a mold to a specific pattern and shape. The titanium fiber metal pads are positioned into recesses in the femoral component substrate and subjected to the diffusion bonding process. The diffusion bonding process employs the use of heat (lower temperatures than sintering) and pressure application to bond the fiber metal wires to the femoral component substrate.[33] The porosity of the titanium fiber metal pads is controlled by the shape and pattern of the wire.

For the plasma-sprayed CPTi coating, the coating material is heated within a spray nozzle. The coating powder (CPTi) and a pressurized gas mixture are then injected into a high-energy arc created within the nozzle and the molten powder is propelled against the implant surface. The application of a plasma spray coating on titanium alloys also diminishes the fatigue performance of the implant due to the increase in sites for crack initiation and propagation. The characteristics of the plasma sprayed coating are controlled through variations in particle size of the CPTi powder and the pressure applied.

In an attempt to improve the fixation of bone to implants, a key research focus has been the development of surfaces that encourage biological fixation. Bioactive ceramic coatings applied to the surfaces of metallic implants have been intensely investigated.[34-50] Theoretically, a biochemical bond forms between bone and a calcium phosphate ceramic coating, typically hydroxyapatite (HA). Improved implant stability is achieved through this biochemical bond. HA has been shown to enhance the interfacial shear strength and bone contact in animal models when plasma sprayed on a titanium alloy (Ti + 6Al + 4V) implant.[34-44] HA has also been shown to negate the effects, specifically fibrous tissue formation, of an initial surgical gap[45-47,49,50] and micromotion between the bone and implant[48] in animal models. A macrotextured surface with a bioactive coating would have the advantages of enhanced early bone formation and the elimination of fibrous tissue formation, resulting in a strong mechanical fixation with the bone.

The purpose of this study was to investigate arc deposition of CPTi, a new implant surface macrotexturing technique, with AD/HA and a plasma-sprayed HA coating without arc deposition (AD). Multiple analysis techniques were utilized including acoustic microscopy, mechanical testing, qualitative histology, quantitative histomorphometry, and scanning electron microscopy.[51] We will only report on the acoustic microscopy results in this chapter.

Ten purpose-bred coonhounds received staged bilateral hemiarthroplasties in the proximal femurs using the unique Harrington canine femoral component model (HARCMOD HIP). One femur received an uncoated Ti + 6Al + 4V alloy femoral stem with an arc-deposited commercially pure titanium (CPTi) surface. The arc deposition process was achieved by striking an electric arc between two pure titanium wires. A high pressure carrier gas passed through the arc and atomized the melt. This resulted in a spray directed at the alloy substrate. Coating build-up resulted from spatter/cooling of matter spray.

The contralateral femur received a similar stem that had a plasma sprayed hydroxyapatite (HA) coating with a nominal thickness of 50 μm over the CPTi surface. The hydroxyapatite was plasma sprayed onto the surface of the metal. The crystallinity was at least 65% and the coating was 90% pure HA. The surface roughness of the CPTi coating was 30 to 35% greater than the HA coating.

Fig. 4.16 shows the CPTi surface and Fig. 4.17 shows the HA-coated CPTi surface. The femurs were retrieved after 6, 12, and 24 weeks of implantation. The implants were sectioned into seven 6-mm-thick cross sections numbered proximally to distally. Section 4 (center of stem) and section 7 (distal stem) were used for SAM analysis. Fig. 4.18 shows the section locations. Sections were dehydrated in ethanol and embedded in polymethylmethacrylate (PMMA) as previously described.

Twenty-four samples were used for SAM. They were evenly distributed between section 4 and section 7, HA and nonHA coatings, and over the 6-, 12-, and 24-week periods. Samples were ground to a 600 grit finish and scanned with a 50 MHz transducer using an 80 μm resolution and a scanned area of 20 mm × 20 mm. The scanning acoustic microscope was calibrated as previously described.

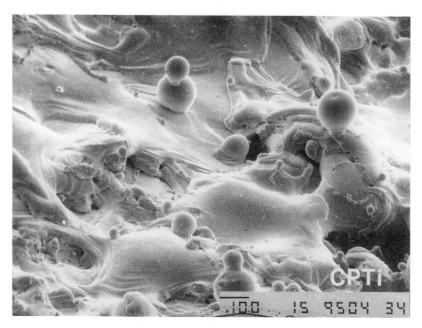

FIGURE 4.16 Secondary electron imaging (SEI) scanning electron photomicrograph of an unimplanted AD surface taken at an angle of 50° in order to show surface topography. Original magnification × 75.

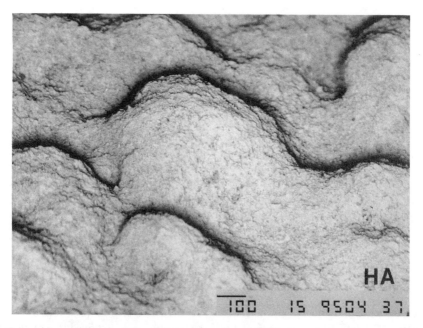

FIGURE 4.17 Secondary electron imaging (SEI) scanning electron photomicrograph of an unimplanted AD/HA surface taken at an angle of 50° in order to show surface topography. Original magnification × 75.

Thresholds were set that eliminated signals from metal and PMMA during analysis. Impedance measurements were taken of (1) the entire sample, (2) the anterior, posterior, medial, and lateral cortical bone regions, and (3) periosteal bone regions when they were present. One-factor analysis of variance (ANOVA) was performed to compare the impedance of the entire sample with respect to implantation time, femoral position of the implant, and surface treatment. A one-factor ANOVA was also performed

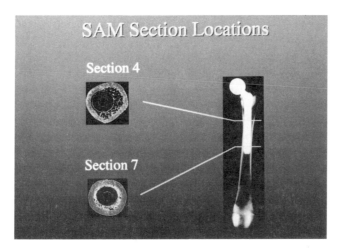

FIGURE 4.18 Schematic implanted specimen indicating sectioning levels for acoustic microscopy.

to test the significance of quadrant location (anterior, posterior, medial, and lateral) on the impedance of cortical bone. A paired t-test was performed to compare the impedance of cortical and corresponding periosteal bone.

The mean acoustic impedance of the canine femoral bone from all of the implant specimens was 8.74 ± 0.25 MRayl. A graph of bone impedances (means) for different implantation times is shown in Fig. 4.19. No statistical difference was found in total impedance between section 4 and section 7 or between HA coated and uncoated CPTi surfaces when analyzing all of the data grouped together. There was a significant difference between the 6-week (8.58 ± 0.35 MRayl) and 24-week (8.87 ± 0.14 MRayl) periods (N = 24, p-value = 0.0214). The difference at the 12-week period (8.77 ± 0.14 MRayl), however, was not statistically significant. A one-factor ANOVA (N = 96) of the cortical bone measured at the four quadrant locations showed a statistically significant difference between posterior (8.82 ± 0.35 MRayl) and anterior (9.04 ± 0.32 MRayl, p-value = 0.0452), posterior and medial (9.06 ± 0.44 MRayl, p-value = 0.0296), and posterior and lateral (9.07 ± 0.38 MRayl, p-value = 0.0208) quadrants. Four of the samples exhibited periosteal bone growth and a paired t-test (N = 4) showed a statistically significant difference between cortical bone impedance (8.62 ± 0.47 MRayl) and periosteal bone impedance (7.09 ± 0.59 MRayl, p-value = 0.0034). Eight transverse cross sections taken from normal canine femora (4 from section 4 and 4 from section 7) had average impedances of 9.34 MRayls.

Fig. 4.20 shows a typical acoustic scan of the canine bone. The implant is located in the center of the medullary canal and trabecular bone struts can be observed bridging the gap between the cortical bone and the implant. This bone has lower acoustic impedance in the range of 6 to 7 MRayls. The histogram highlights the range of properties. The cortical bone has relatively high impedance (greater than 9 MRayls) and there is evidence of remodeling in the posterior quadrant as well as the endosteal area of the bone.

The results of this study demonstrated a significant increase in canine femoral bone impedance around a Ti stem between 6 and 24 weeks post-implantation, suggesting a greater bone stiffness at 24 weeks. This experiment also demonstrated that the acoustic properties of canine femoral bone after implantation of a hip stem differ by position within the bone. The impedance of the posterior region was shown to be less than that of the anterior, medial, and lateral regions. This latter finding is consistent with the work of Ashman et al.[26] who demonstrated a decrease in C_{33} and density in the posterior regions of human and canine femurs as compared with the other quadrants of the bone. No differences in total sample impedance or sample quadrant impedance were observed when comparing samples from section 4 with those of section 7. Moreover, no difference in impedance was shown between HA and uncoated CPTi specimens.

Fig. 4.21 highlights changes in the cortical thickness of the same canine femora that were scanned. A significant reducton in cortical thickness was observed while the average impedance increased. This is a very interesting result that demonstrates how bone remodeling can result in a net decrease in material

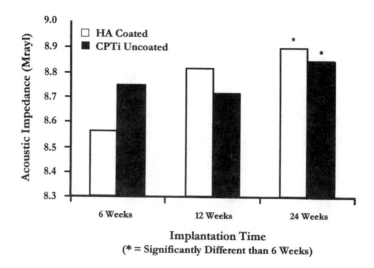

FIGURE 4.19 Graph showing impedance vs. implantation times for CPTi and HaCPTi implants.

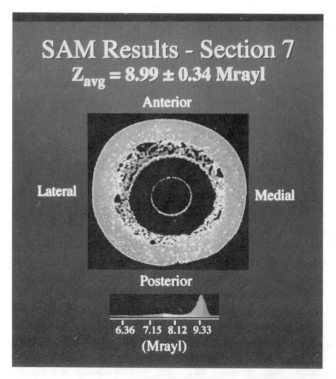

FIGURE 4.20 Typical acoustic scan for a CPTi implant with trabecular bone struts bonding the implant to the bone. The reduced impedance properties of the bone struts are obvious with the cortical bone having more mineralized bone with greater impedance.

yet the properties of the material can remain the same or increase. While cortical thickness reflects a change in volume, the increase in impedance reflects increased calcification and maturation of the bone. This result can be confusing because new intramedullary bone formed within the canal and adjacent to the implant would certainly mature with time and would also affect the average impedance of the section.

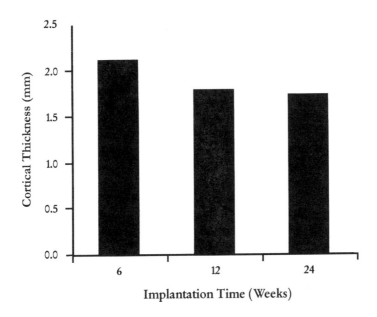

FIGURE 4.21 Mean cortical bone thickness at 6, 12, and 24 weeks for lumped data for AD/HA and AD implants.

The Analysis of Bone Remodeling Adjacent to Absorbable Polymers

Synthetic degradable polymers are currently under investigation for orthopaedic applications ranging from small bone fixation pins to scaffolds for tissue engineered bone regeneration. While traditional histological, microscopic, and microradiographic techniques yield important insights into the host bone response, they are not able to reveal the effects the polymers may have on the mechanical properties of the bone local to the implant site. In this experiment, scanning acoustic microscopy was used to nondestructively and quantitatively assess the micromechanical properties of bone growing adjacent to poly(L-lactic acid) (PLA) and tyrosine-derived polycarbonate implants.[52]

PLA is biocompatible in the short term, and able to provide adequate strength and stiffness for use in small bone fixation.[53-55] However, PLA implants have been associated with late inflammatory and bone resorption responses.[55] Concerns have been raised regarding the acidic nature of the degradation products and the particulate debris formed during erosion.[56,57] Tyrosine-derived polycarbonates are part of a newly synthesized class of degradable polymers.[58-60] Like PLA, these materials are based on natural metabolites (the amino acid, tyrosine), have favorable mechanical properties, are readily processible, and degrade on the order of months to years. Unlike PLA, tyrosine-derived polycarbonates are completely amorphous. Hence, degradation to crystalline particulate debris is not of practical concern for these materials. Similarly, the degradation products of tyrosine-derived polycarbonates are nonacidic. *In vitro* cytotoxicity[58] and short-term *in vivo* evaluations in rats[61] and rabbits[62] have shown tyrosine-derived polycarbonates to be generally biocompatible. Poly(DTE carbonate) (molecular weight = 121,000) and poly(DTH carbonate) (molecular weight = 220,000) were synthesized according to previously published procedures.[54] Medical grade PLA (molecular weight = 191,000) was obtained from Boehringer Ingelheim in pellet form and used as delivered.

The canine bone chamber model employed in this study and corresponding histological and radiographic findings have been previously described.[63] Briefly, compression molded coupons were fit into a polyethylene bone chamber housing to create 10 ingrowth channels $1 \times 5 \times 10$ mm. Assembled chambers were implanted in a cortical defect created in the lateral metaphysis of the distal femur. Chambers from three dogs at 6-, 12-, 24-, and 48-week time points were analyzed. This yielded two chambers per polymer per time point. The retrieved chambers were fixed, imbedded, and polished as previously described. The specimens were cut perpendicularly through the centers of the bone chambers and the exposed faces polished for SAM analysis.

The mean impedance of the bone that had penetrated the channels of the bone chambers is summarized in Fig. 4.22. No appreciable difference in the properties of bone growing adjacent to the different polymers was noted at the first three time periods (6, 12, and 24 weeks). All chambers demonstrated good bone ingrowth and a trend toward increasing impedance. However, 48 weeks post-implantation there was a marked decrease in the impedance of the bone in the PLA chambers in comparison to the 24-week PLA data. This was in direct contrast to the bone growing near the tyrosine-derived polymers which was still increasing in impedance at 48 weeks.

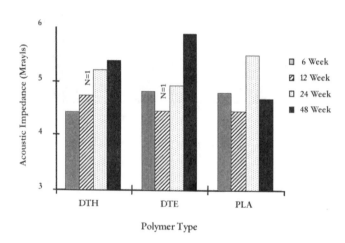

FIGURE 4.22 Bone chamber impedance for poly DTE, poly DTH, and PLA from canine femoral specimens at 6, 12, 24, and 48 weeks.

Figs. 4.23 and 4.24 show acoustic scans of 48-week bone chambers for PLA and PolyDTE, respectively. Very little bone is observed growing into the PLA chamber and the impedance is significantly lower than the impedance at 48 weeks for polyDTE. In contrast, well-mineralized bone is observed filling the channels of the DTE bone chamber.

Using acoustic microscopy, a quantitative micromechanical analysis of bone growing adjacent to biodegradable polymers was possible. The decrease in the elastic properties of the bone juxtaposed to PLA from 24 to 48 weeks possibly reflects the effect of PLA degradation products. Hypothetically, PLA degradation results in a more acidic local environment, causing bone resorption (seen histologically) and consequently a reduction in the impedance of the surrounding bone. On the other hand, bone growing into the tyrosine-derived polycarbonate implants continued to calcify out to 48 weeks and approached 75% of the impedance for normal canine cortical bone.

4.4   Acoustic Microscopy of Partially Mineralized Tissues

Problems and Obstacles in the Mechanical Characterization of Incompletely Calcified Tissues

Classically, the mechanical behavior of bone has been determined for consistent and relatively homogeneous specimens machined or otherwise shaped from diaphyseal compact bone.[64-66] Such samples are typically subjected to tensile or compression tests. This methodology works quite well for the characterization of this type of bony specimen. However, the properties thusly determined are by definition averaged over the specimen's entire cross section and length. Similar testing protocols have been implemented for cores of cancellous bone[67,68] and individual trabeculae.[69,70] While this research provides useful

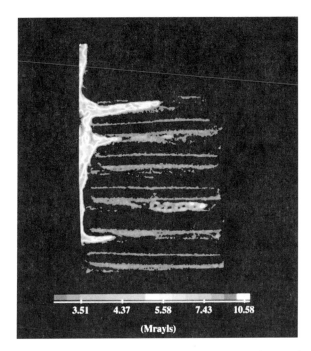

FIGURE 4.23 Typical acoustic scan for a 48-week PLA implant. Relatively little bone is found within the chamber and the impedance is significantly lower.

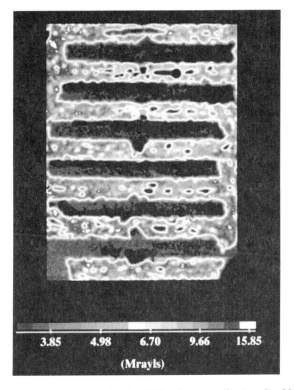

FIGURE 4.24 Typical acoustic scan for a 48-week polyDTE implant. A well-mineralized bone is found adjacent to the polymer coupons.

and important information, these techniques are inherently unsuitable for the investigation of bone formed in rapidly remodeling environments.

The tissue in such instances is highly heterogeneous and anisotropic. Conventional testing procedures that measure bulk specimen properties cannot easily deal with this type of specimen. Bone has a very irregular and delicate structure, making the fabrication and testing of specimens difficult. Numerous researchers have attempted to overcome these obstacles using other techniques. Spatial variations in material hardness can be investigated using a host of indentation tests (Vickers, Brinell, and Rockwell). Each technique uses indentors of a particular size and geometry, from which a measure of material stiffness can be determined. Several investigators[71,72] have successfully used microindentation methods to assess local property variations in various well-mineralized tissues.

Additional research efforts have employed similar methods toward the characterization of local property variations in partially mineralized bone and fracture callus.[21,73,74] While providing information with much greater detail than conventional mechanical testing, the information is gained only for the particular point sampled. The spacing of indents must be rather large due to the effect of each indent on the surrounding area since the material is damaged (work-hardened) for some distance from the point of indentation. The response of the material at the indent site is also dependent on the properties of the adjacent material buttressing the deformed material. Indentation techniques are not well suited for heterogeneous materials. They are also time consuming and cause permanent damage to the specimen.

The mineral density of irregularly shaped and partially calcified bone can also be determined using dual X-ray energy absorptiometry (DEXA). This can be done *in vivo*; however, the resolving capability of most systems is relatively low (approximately 100 μm), and interpreting the results is difficult due to averaging across specimen thickness. Recent advances in micro-computer tomography (micro-CT) have made it possible to evaluate mineralized tissues with complex and irregular geometries with remarkable precision.[75] These systems have been used to reproduce the architecture of cancellous bone as well as lengthened bone segments.

Advantages of Acoustic Microscopy

Using acoustic microscopy one can quickly and easily generate an image in which the contrast is based solely on differing elastic properties, taking into account both material structure and mineral density. When considering calcified tissues in a dynamic state, scanning acoustic microscopy provides numerous advantages. Since it is a graphic technique, it is well suited for evaluating material with a wide range of properties. The gradations and trends in elastic properties are readily apparent throughout the images. Additionally, no specimen processing is necessary other than embedding and sectioning, so an *in situ* evaluation of the tissue properties can be attained. Measurement of acoustic properties in irregularly shaped areas is another great strength of this technology.

In contrast to micro-CT and DEXA, the images generated using scanning acoustic microscopy (SAM) are based on both mineral density and tissue ultrastructure. Therefore, the images are representative of the elastic behavior of the bone. Additionally, a much greater resolution is possible using SAM, and the acoustic information gathered constitutes a surface property measurement, not an average over a thickness as is the case with the volume elements used in micro-CT. Alone or in conjunction with other methodologies, SAM provides a unique way to gain insight into the development of dynamically remodeling environments.

Regional and Temporal Changes in the Acoustic Properties of Fracture Callus in Secondary Bone Healing

In secondary bone healing, the fracture callus provides the limb with temporary stability until the fracture site is replaced with new bone. This process of callus formation involves both intramembranous and endochondral ossification. These two processes occur at different locations and times during the healing process and hypothetically have distinct roles in the restoration of mechanical competence to the skeleton. Thus, the optimum manner in which to evaluate the mechanical properties of fracture callus is not a clear-cut issue. This is of great concern to the orthopedic investigator because a reliable measure of the mechanical properties of healing bones often plays a pivotal role in the success or failure of many fracture healing studies.

The purpose of this study was to characterize the acoustic properties of healing fractures in normal rats using a scanning acoustic microscope.[76] Unilateral, transverse, and mid-diaphyseal fractures of the right femur were created in 17 Sprague-Dawley rats weighing approximately 400 gm. The animals were sacrificed at 2 (n = 4), 4 (n = 5), 6 (n = 4), and 8 (n = 4) weeks after fracture. Contralateral left femurs were collected from five rats to serve as a control group. The limbs were fixed in 10% buffered formalin and embedded in polymethylmethacrylate (PMMA). Mid-sagittal sections were prepared and analyzed using acoustic microscope.

Each acoustic image was partitioned into six non-overlapping regions, three on the anterior surface and three on the posterior surface as depicted in Fig. 4.25. This was done in an attempt to isolate tissue formed through endochondral and intramembranous pathways without including the original cortical bone. The acoustic impedance and area of bone in each of these regions were then measured using our custom software. The normal limbs were analyzed in a similar fashion to determine the properties of intact bone. However, since no fractures were present in the normal bones, the measurements were made by centering the middle region mid-shaft on the femur.

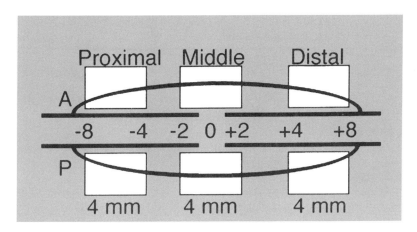

FIGURE 4.25 Idealized fracture callus divided into six regions for the measurement of acoustic properties.

The six regions measured on each scan were used to compute a mean impedance for the fracture callus of each specimen. This was accomplished by calculating a weighted average based on the area of bone present in each of the six regions. Similarly, the mean impedance of the callus in the proximal, middle, and distal locations was determined by computing a weighted average of the two corresponding regions on the anterior and posterior aspects of each limb. These three locations (proximal, middle, and distal) were compared at each time period to determine the regional variations within the callus. The variations occurring within each of these locations with respect to healing time were also examined in order to track the development of bone from specific regions. A single-factor ANOVA was used to analyze the mean impedance of the entire callus at each time period, while a two-factor ANOVA with repeated measures was used to examine the data from the three different regions. When differences were indicated, Bonferroni-Dunn *post hoc* tests were performed at the 95% confidence level to find the significant comparisons.

A significant difference (p < 0.0013 for all comparisons) was observed in the mean impedance of the fracture callus from each of the five experimental groups, including the normal femora. Furthermore, the data indicated that an extremely linear relationship (r^2 = 0.999) between healing time and mean callus impedance existed as seen in Fig. 4.26. Fig. 4.27 details the regional variations in callus impedance

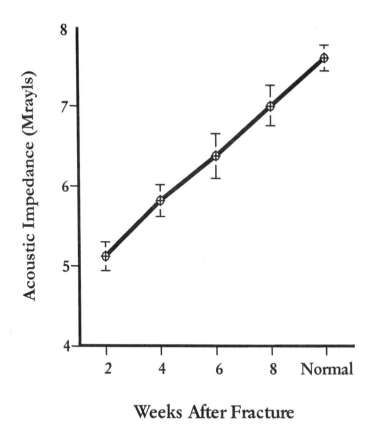

FIGURE 4.26 Mean impedance of fracture callus vs. healing time, demonstrating a nearly linear relationship ($r^2 =$ 0.999).

at each time interval. Two weeks after fracture the middle region (4.90 ± 0.13, mean impedance ± SD) was of significantly lower impedance ($p < 0.0023$) than the proximal region (5.47 ± 0.19). Four weeks after fracture all locations were different, with the proximal (6.59 ± 0.32) having the greatest impedance and the middle (5.44 ± 0.20) having the least. Six weeks post-fracture the proximal callus (7.15 ± 0.35) still exhibited higher impedance values than the middle (6.00 ± 0.30, $p < 0.0017$) and distal regions (6.79 ± 0.13, $p < 0.0106$). However, the middle and distal regions were statistically equal at that time. Eight weeks after fracture, all three regions of the callus were statistically equivalent. Interestingly, the bones of the normal femora exhibited significantly different impedances, $p < 0.012$, in the proximal (7.97 ± 0.18), middle (7.54 ± 0.06), and distal (7.13 ± 0.33) locations.

Evaluating the data by comparing the variations occurring within each region (proximal, middle, and distal) with respect to healing time showed that by 8 weeks post-fracture the impedance of the middle region of the callus was statistically equivalent to the impedance of bone in the middle region (mid-diaphysis) of normal femora. However, at 8 weeks the proximal and distal regions of callus were statistically different from their corresponding regions of normal bone, $p \leq 0.0012$ proximal and $p \leq 0.0021$ distal. Additionally, the 4- and 6-week impedances were found to be statistically equivalent within similar regions, i.e., the proximal region was statistically equal between 4 and 6 weeks.

In the 2-week specimens, no bridging callus was observed using the SAM. The mineralization fronts seen then were of consistently lower impedance than the intramembranous callus. As seen in Fig. 4.28, bridging callus was present in all the specimens in the 4-week group. A broad range of acoustic properties, as demonstrated by the foci of high impedance in Fig. 4.28, was characteristically noted in the middle region of the callus 4 weeks post-fracture. In contrast, the acoustic properties were much more consistent in the callus located further away from the fracture site. Six weeks post-fracture, the acoustic properties

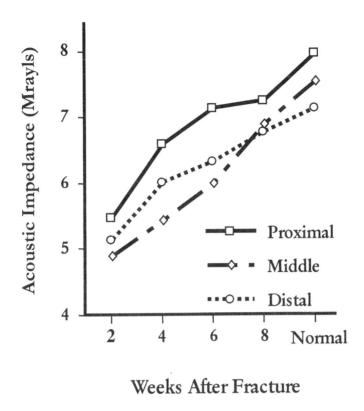

FIGURE 4.27 Acoustic impedance of the various regions (proximal, middle, and distal) within the callus vs. healing time.

of the callus had become much more consistent as remodeling progressed, as demonstrated in Fig. 4.29. The callus of the 8-week group was marked by an even greater uniformity in callus properties and at times was difficult to discern from the bone in the original cortices.

In general, the time course of increasing impedance in the present study is consistent with the changes in the calcium content of fracture callus reported previously in rats[73,77] and in dogs.[74] Although we observed significant differences in the impedance of the callus at each time period, Aro et al.[73] observed that mineralization of the external callus did not change significantly between 3 and 6 weeks post-fracture. In the same study, a similar trend was seen in callus hardness as measured by microindentation. Hardness increased slowly and insignificantly up to 3 weeks, sharply increased between weeks 3 and 4, then did not change between weeks 4 and 6. The model used in that study, however, consisted of bilateral tibial fractures and 0.4 mm stainless steel nails for stabilization. These factors may account for some of the discrepancies with the results presented here.

Millett et al.[78] measured bone mineral density (BMD) in a model identical to the one used in the present study. BMD was measured with a DEXA scanner and an ultrahigh-resolution software package. A significant increase between 2 and 8 weeks but not between 8 and 12 weeks was reported. The mean increase in BMD between 2 and 8 weeks for all sections analyzed was 12.4%. In the present study, the mean increase in acoustic impedance between 2 and 8 weeks in similar regions was 37%. The rather large difference in these two measurements may be a result of the structural changes detected by acoustic microscopy. Significant increases in tissue organization will occur as the fracture site remodels. These changes, as well as mineral density increases, may explain the relatively larger impedance increases seen with the SAM.

Walsh et al.[79] examined the mechanical properties of unilateral fractures in the femora of rats stabilized with 0.45 mm K-wires. A rapid rise in peak tensile load was noted, although 6 weeks after fracture the

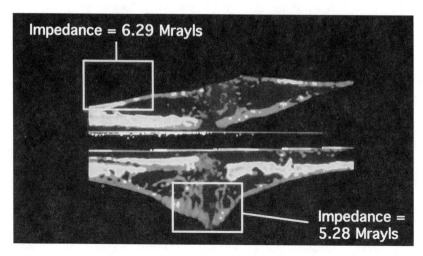

FIGURE 4.28 Acoustic image of a rat femur 4 weeks post-fracture. Boxes indicated foci of high impedance in the callus, illustrating the highly heterogeneous distribution of acoustic properties characteristically seen at this time period.

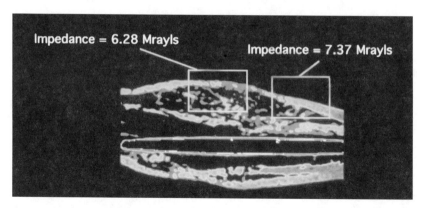

FIGURE 4.29 Acoustic image of a rat femur 6 weeks post-fracture. The formation of a new cortical shell with a wide range of acoustic properties was typically present in the specimens at this time period.

experimental limbs demonstrated only 50% of the tensile force to failure of their contralateral normals. Variables in all of these studies include bilateral vs. unilateral models, presence and size of K-wires for fixation, and the ages and sizes of the animals. To be sure, these factors will have some effect on the outcome, and make direct comparisons between studies difficult. Overall, the relationship between acoustic impedance and healing time seen in the present study was more linear, and had less variation within each time period than any mechanical or physical parameters reported in the literature for a fracture healing model. It is possible that the acoustic properties of these tissues may provide a more representative measure of their development than any previously used techniques.

A unique feature of this study is that the acoustic trends we observed in the different regions may be interpreted based on our histological findings. For example, the impedance of the middle region did not begin to rapidly increase until 4 weeks post-fracture. Histologically, this corresponded to the time period when the middle region was beginning to mineralize. Six weeks post-fracture the impedance of this region (middle) was still rapidly increasing and histologically we observed a much more remodeled callus with the development of woven bone. Thus, by using SAM, it is possible to get some idea of how the various reparative processes contribute to the development of elastic properties within the callus. Since the stability and ultimate healing of fractured bones will be functions of the stiffness of the callus immediately adjacent to the fracture site, the determination of the acoustic properties in this region may provide a more realistic measure of fracture healing.

In summary, the data in this experiment indicated that during the early stages of fracture healing, callus formed through intramembranous ossification has a more rapid rise in impedance than endochondral callus. However, as the transformation to woven bone takes place, the acoustic properties of the middle portion of the callus (endochondral origin) quickly increased to equal the proximal and distal regions by the eighth week post-fracture. Bronk et al.[80] hypothesized that different mechanical test results (tension vs. torsion) are affected by different healing events. The data presented here may provide indirect evidence in support of their findings.

The ability of the SAM technique to accurately image local acoustic property gradients can be a great strength when studying highly heterogeneous and dynamic environments such as those encountered during fracture healing. As demonstrated in this study, we were able to quantitatively measure variations in the callus from different locations with a precision that to date has not been accomplished. While these results in and of themselves are significant, the proven ability to image and measure these variations may have more widespread implications. Many factors thought to enhance the process of fracture repair, such as growth factors, micromotion, or electromagnetic fields, are likely to have only regionally specific effects which might only be detected with acoustic microscopy.

The Acoustic Properties of Bone Formed during Limb Lengthening

Limb lengthening as the name implies, refers to the clinical practice of increasing the length of the long bones of the skeleton. In contemporary orthopedics, this is nearly synonymous with the process of distraction osteogenesis. This technique was first developed and practiced during the 1950s by Dr. Gavril Ilizarov in Sibera. Briefly, this process involves the steady and slow separation of two bone fragments after the surgical creation of a fracture. An external fixation device is used to provide stability, and a means for gradually increasing the distance between the bone ends via an adjustable mechanism. The actual process of distraction or lengthening ensues after a latency period of a few days to allow for initial healing and callus formation. The new tissue (or bone regenerate) formed in the created gap begins to mineralize from the original bone ends toward the center of the gap. Many factors affect the outcomes of these procedures, and for that reason distraction osteogenesis has become an active area of research in the orthopedic community.

Assessing the mechanical properties of the bone formed during distraction osteogenesis is not an easy task. At the end of distraction, the morphology of the regenerate consists of axially aligned cones, or pyramidal-like projections of bone extending toward an unmineralized fibrous interzone at the center of the gap. The width of this interzone gradually decreases as the regenerate matures. Evaluating the properties of the regenerate is complicated for many reasons. During the early stages of regeneration, there are wide ranges in mineral content, structure, and location. In other words, as seen during fracture healing, bone is a heterogeneous, irregular, and anisotropic tissue. Analyses characteristically are conducted using microradiography and/or various histological techniques, most of which are cumbersome. With an acoustic microscope, it is possible to quantify how the various factors affect the material properties of the new bone, as well as monitor the maturation process. A series of experiments were undertaken to examine the effects of several variables on the acoustic properties of bone formed during distraction osteogenesis.

The first experiment[81] was designed to evaluate the effects of lengthening rate. Unilateral mid-diaphyseal limb lengthenings were performed on the right tibiae of eight skeletally mature, male, New Zealand white rabbits. An Orthofix M-100 mini-lengthener was applied to the anteromedial aspect of each limb. Four animals were lengthened at a rate of 1.4 mm/day, while another four were lengthened at 0.70 mm/day. Half of the animals in each group were lengthened to achieve a maximum length of 14 mm. The remaining animals in each group were sacrificed midway through the lengthening process. The animals from the slower distraction group (0.70 mm/day) had formed new bone with significantly greater (p < 0.05) acoustic properties than those subjected to the faster (1.4 mm/day) rate of distraction.

Interestingly, the volume of callus formed was similar between the two groups. It appeared that the slower rate of distraction allowed the bone regenerate to mature more rapidly, although there was no net increase in new bone deposition. A previous study investigating lengthening rates in an identical model concluded that there was no difference in bone formation using the same two rates of distraction.[82] This finding was most likely the result of using qualitative analysis techniques that were unable to discern distinct differences between the experimental groups.

The hypothesis of the second study[83] was that the preservation of the internal blood supply via a corticotomy technique would establish a more favorable environment for limb lengthening, resulting in bone regenerate with greater acoustic properties. Previous studies investigating the importance of preserving the medullary vasculature during distraction osteogenesis have generated conflicting and/or inconclusive findings.[84-86] As in the first experiment, lengthenings were performed on the right tibiae of skeletally mature New Zealand white rabbits using the same hardware and techniques. An osteotomy was performed using one of two techniques (day 0). In Group 1 (18 animals), a simple osteotomy was performed using an air driven oscillating saw. Group 2, consisting of an additional 18 animals received corticotomies. The corticotomy consisted of perforating the tibia with a 1.1 mm drill around its circumference without entering the medullary space. The fracture was completed by gently bending the limb with the fixator loosely applied to prevent excessive displacement of the bone ends.

Ten days after surgery, lengthening was started at a rate of 0.35mm/12h for a period of 20 days in both groups. Weekly radiographs were taken to monitor the lengthening process. At the end of distraction (Day 30), six animals from Group 1 and six animals from Group 2 were sacrificed. Three weeks later (Day 51), six animals from Group 1 and six from Group 2 were sacrificed. The remaining animals were sacrificed three weeks later (Day 72).

As in the previous studies, the limbs were fixed in formalin and embedded in PMMA. The tibiae were sectioned in the sagittal plane, and prepared for evaluation with a scanning acoustic microscope (SAM). Only material located directly in the lengthened gap was included in this analysis to avoid the effects of periosteal callus and original cortical bone. This process was also modified to exclude the effect of the embedding material. Thus, only the impedance of the bone regenerate was determined. The data were analyzed using t-tests and a 0.05% significance level. Comparisons were made between osteotomy methods at each time period and between the three time periods as well.

The mean impedance of the bone regenerate at the end of distraction was 3.74 ± 0.07 MRayls for the corticotomized animals and 3.64 ± 0.03 MRayls for the osteotomized group. These two groups were significantly different ($p = 0.016$). Three weeks post-distraction, the mean impedance of the corticotomized and osteotomized groups was 4.19 ± 0.19 and 4.07 ± 0.64 MRayls, respectively. Three weeks post-distraction, both groups were statistically equivalent. Six weeks post-distraction, the impedance of the corticotomized group was 4.81 ± 0.28, whereas, the osteotomized group impedance was 4.65 ± 0.33. As in the 3-week group, there was not a significant difference in the data at 6 weeks. Despite the lack of statistical significance at 3 and 6 weeks, it is worthwhile to note that the acoustic impedance of the regenerate in the corticotomized series was consistently greater. Pooling the data at each time point showed that the impedance of the regenerate at 3 weeks post-distraction was significantly greater than at the end of distraction ($p < 0.0001$). As a point of reference, the impedance of the normal rabbit tibial cortical bone was observed to consistently have values of approximately 8 MRayls. These results are summarized in Fig. 4.30.

Fig. 4.31 presents typical acoustic images of a limb lengthened corticotomy and osteotomy at the end of lengthening. The corticotomy demonstrates advanced healing with an average acoustic impedance of 4.05 MRayls. The osteotomy has an average impedance that is significantly lower at 3.47 MRayls.

A major shortcoming of most limb lengthening studies is their inability to quantitatively assess the effect of the factors under investigation and the quality of the new bone formed. In the present study, we have quantitatively detected differences in the material properties of the bone regenerate formed as a result of two different corticotomy methods. This is a highly significant finding in limb lengthening research. Previous studies have concluded that complete severing of the endosteal blood supply has little or no impact on distraction osteogenesis.[84-86] The similarity between the two groups at 3 and 6 weeks

Weeks Post Lengthening	Corticotomy	Osteotomy
0*	3.75 ± 0.07	3.64 ± 0.03
3	4.19 ± 0.19	4.08 ± 0.64
6	4.81 ± 0.28	4.65 ± 0.33

* ($p = 0.016$)

FIGURE 4.30 Table of mean acoustic impedances for limb lengthened corticotomies and osteotomies at 0, 3, and 6 weeks post-lengthening.

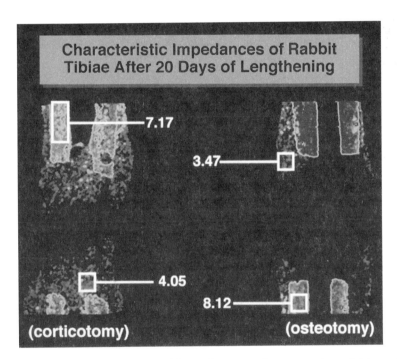

FIGURE 4.31 Acoustic scan of a limb lengthened by corticotomy and osteotomy immediately post-lengthening. Regional variations in impedance are highlighted. The corticotomy callus has greater impedance than the osteotomy callus (4.05 vs. 3.47 Mrayls).

post-distraction is not surprising because the blood supply has most likely been fully restored at this time in the osteotomized animals. The study of regenerate properties at later time periods may yield a more accurate description of the biomechanical development of the bone regenerate than is possible with conventional mechanical testing techniques.

Acknowledgments

We would like to acknowledge the following individuals for their contributions to this ongoing scientific endeavor: Hubert Berndt, Matt Walenciak, Sophie Weiss, Frank Alberta, Biren Chokshi, Navine Budwani, Avinash Prabhakar, Eli Hurowitz, Raj Arakal, Grace Chowchuvech, Tom Poandl, J. Russell Parsons, Ph.D.,

Francis Lee, M.D., Ph.D., Bob Gilmore, Ph.D., John Young, Ph.D., Ken James, Ph.D., Jack Ricci, Ph.D., Joachim Kohn, Ph.D., Alex DiPaula, and Nejat Guzelsu, Ph.D.

The following organizations have contributed to the funding of these projects: The Focused Giving Program of the Robert Wood Johnson Foundation, The Orthopaedic Research and Education Foundation, General Electric Inc., The Whitaker Foundation, The Fulbright Foundation, Osteonics, Inc., Orthofix Ltd., and the Department of Orthopaedics of the New Jersey Medical School.

References

1. Kolsky, H., *Stress Waves in Solids*, Clarendon Press, Oxford, 1953.
2. Beyer, R. and Lechter, S., *Physical Ultrasonics*, Academic Press, New York, 1969.
3. Lemons, R. and Quate, C., *Appl. Phys. Lett.*, 24, 163, 1974.
4. Jipson, V. and Quate, C., *Appl. Phys. Lett.*, 32, 789, 1978.
5. Gilmore, R., Joynson, R., Trzaskos, C., and Young, J., *Phil. Trans. R. Soc.*, 320, 215, 1986.
6. Kessler, L., *Metals Handbook*, 9th ed., ASME International, Materials Park, OH, 1989, Vol. 17, 465.
7. Bereiter-Hahn, J. and Buhles, N., *Imaging and Visual Documentation in Medicine*, Elsevier, Amsterdam, 1987, 537.
8. Bereiter-Hahn, J., *Scanning Image Technol.*, 809, 162, 1987.
9. Bereiter-Hahn, J., Litniewski, J., Hillmann, K., Krapohl, A., and Zylberg, L., *Acoustical Imaging*, Plenum Press, New York, 1989, 27.
10. Briggs, G., Daft, C., Fagan, A., Field, T., Lawrence, C., Montoto, M., Peck, S., Rodriguez, A., and Scruby, C., *Acoustical Imaging*, Plenum Press, New York, 1989, 1.
11. Lees, S., Tao, N.J., and Lindsay, S.M., *Acoustical Imaging*, Plenum Press, New York, 1989, 371.
12. Okawai, H., Tanaka, M., Dunn, F., Chubachi, N., and Honda, K., *Acoustical Imaging*, Plenum Press, New York, 1989, 193.
13. Kasahara, S., Yoshida, K., Kushibiki, J., and Chubachi, N., *Acoustical Imaging*, Plenum Press, New York, 1989, 153.
14. Meunier, A., Katz, J.L., Kristal, P., and Sedel, L., *J. Orthop. Res.*, 6, 770, 1988.
15. Tatento, H., Iwashita, Y., Kawano, K., and Noikura, T., *Acoustical Imaging*, Plenum Press, New York, 1989, 125.
16. Zimmerman, M., Meunier, A., Kristal, P., Katz, J., and Sedel, L., *J. Orthop. Res.* 7, 607, 1989a.
17. Zimmerman, M., Meunier, A., Katz, J., and Kristal, P., *IEEE Trans. Biomed. Eng.*, 37, 433, 1990.
18. Katz, J. and Meunier, A., *J. Biomechanical Eng.*, 115, 543, 1993.
19. Gardner, T., Elliott, J., Sklar, Z., and Briggs, G., *J. Biomechanics*, 25, 1265, 1992.
20. Hasegawa, K., Turner, C., Recker, R., Wu, E., and Burr, D., *Bone*, 16, 85, 1995.
21. Broz, J., Simske, S., and Greenberg, A., *J. Biomechanics*, 28, 1357, 1995.
22. Turner, C., Chandran, A., and Pidaparti, V., *Bone*, 17, 85, 1995.
23. Pidaparti, V., Chandran, A., Takano, Y., and Turner, C., *J. Biomechanics*, 29, 909, 1996.
24. Briggs, G., *Acoustic Microscopy*, Oxford University Press, New York, 1992.
25. Zimmerman, M.C., Prabhakar, A., Chokshi, B.V., Budhwani, N., and Berndt, H., *J. Biomed. Mater. Res.*, 28, 931, 1994.
26. Ashman, R., Cowin, S., Van Buskirk, W., and Rice, J., *J. Biomechanics*, 17, 349, 1984.
27. Bloebaum, R.D., Ota, D.T., Skedros, J.G., and Mantas, J.P., *J. Biomed. Mater. Res.*, 27, 1149, 1993.
28. Meunier, A., Riot, O., Kristal, P., and Katz, J.L., *Interface in Medicine and Mechanics*, Elsevier, London, 1991, 454.
29. Gillespy, T., III and Gillespy, M.P., *Rad. Clin. North America*, 29, 77, 1991.
30. Yoon, H. and Katz, J., *J. Biomechanics*, 9, 407, 1976.
31. Harten, R. Jr., Depaula, C., Kotha, S., Zimmerman, M., Parsons, J., and Guzelsu, N., *Transactions of the 23rd Annual Meeting of the Society for Biomaterials*, New Orleans, 1997.
32. Pilliar, R., *Clin. Orthop.*, 176, 42, 1983.
33. Galante, J., Rostoker, W., Lueck, R., and Ray, R., *J. Bone Jt. Surg.*, 53A, 101, 1971.

34. Thomas, K., Kay, J., Cook, S., and Jarcho, M., *J. Biomed. Mater. Res.*, 21, 1395, 1987.
35. Geesink. R., de Groot, K., and Klein, C., *J. Bone Jt. Surg.*, 70B, 17, 1988.
36. Rivero, D., Fox, J., Skipor, A., Urban, R., and Galante, J., *J. Biomed. Mater. Res.*, 22, 191, 1988.
37. Klein, C., Patka, P., van der Lubbe, J., Wolke, J., and de Groot, K., *J. Biomed. Mater. Res.*, 25, 53, 1991.
38. Dhert, W., Klein, C., Jansen, J., van der Velde, E., Vriesde, R., Rozing, P., and de Groot, K., *Biomed. Mater. Res.*, 27, 127, 1993.
39. Dhert, W., Klein, C., Jansen, J., van der Velde, E., Vriesde, R., Rozing, P., and de Groot, K., *Biomed. Mater. Res.*, 25, 1183, 1991.
40. Bauer, T., Gaisser, D., Uratsugi, M., and Reger, S., *Handbook of Bioactive Ceramics*, CRC Press, Boca Raton, FL, 1990, Vol. 2.
41. Hayashi, K., Inadome, T., Mashima, T., and Sugioka, Y., *Biomed. Mater. Res.*, 27, 557, 1993.
42. Jansen, J., van der Waerdan, J., Wolke, J., and de Groot, K., *Biomed. Mater. Res.*, 25, 973, 1991.
43. Jansen, J., van der Waerdan, J., and Wolke, J., *J. Biomed. Mater. Res.*, 27, 603, 1993.
44. Jansen, J., van der Waerdan, J., and Wolke, J., *J. Appl. Biomaterials*, 4, 213, 1993.
45. Carlsson, L., Rostlund, T., Albrektsson, B., and Albrektsson, T., *Acta Orthop. Scand.*, 59, 272, 1988.
46. Soballe, K., Hansen, E., Brockstedt-Rasmussen, H., Penderson, C., and Bunger, C., *Acta Orthop. Scand.*, 61, 299, 1990.
47. Soballe, K., Hansen, E., Brockstedt-Rasmussen, H., Hjortdal, V., Juhl, G., Penderson, C., Hvid, I., and Bunger, C., *Clin. Orthop. Rel. Res.*, 272, 300, 1991.
48. Soballe, K., Hansen, E., Brockstedt-Rasmussen, H., Jorgensen, P., and Bunger, C., *J. Orthop. Res.*, 10, 285, 1992.
49. Maxian, S., Zawadsky, J., and Dunn, M., *J. Biomed. Mater. Res.*, 28, 1311, 1994.
50. Dalton, J., Cook, S., Thomas, K., and Kay, J., *J. Bone Jt. Surg.*, 77A, 97, 1995.
51. Walenciak, M., Zimmerman, M., Harten, R. Jr., Ricci, J., and Stamer, *J. Biomed. Mater. Res.*, 31, 465, 1995.
52. Harten, R. Jr., James, K., Charvet, J., and Zimmerman, M., *Transactions of the 23rd Annual Meeting of the Society for Biomaterials*, New Orleans, 1997.
53. Bostman, O., *J. Bone Jt. Surg.*, 73, 148, 1991.
54. Bucholz, R., Henry, S., and Henley, M., *J. Bone Jt. Surg.*, 76, 319, 1994.
55. Suganuma, J. and Alexander, H., *J. Appl. Biomaterials*, 4, 13, 1993.
56. Taylor, M., Daniels, A., Andriano, K., and Heller, J., *J. Appl. Biomaterials*, 5, 151, 1994.
57. Daniels, A., Taylor, M., Andriano, K., and Heller, J., *Proc. Orthop. Res. Soc.*, 17, 88, 1992.
58. Ertel, S. and Kohn, J., *J. Biomed. Mater. Res.*, 28, 919, 1994.
59. Kohn, J., *Drug New Perspect.*, 4, 289, 1991.
60. Pulapura, S. and Kohn, J., *Biopolymers*, 32, 411, 1992.
61. Silver, F., Marks, M., Kato, Y., Li, C., Pulapura, S., and Kohn, J., *Long-Term Effects. Med. Implants*, 1, 329, 1992.
62. Ertel, S., Kohn, J., Zimmerman, M., and Parsons, J., *J. Biomed. Mater. Res.*, 29, 1337, 1995.
63. Choueka, J., Charvet, J., Koval, K., Alexander, H., James, K., Hooper, K., and Kohn, J., *J. Biomed. Mater. Res.*, 31, 35, 1996.
64. Keller, T., Mao, Z., and Spengler, D., *J. Orthop. Res.*, 8, 592, 1990.
65. Reilly, D., Burstein, A., and Frankel, V., *J. Biomechanics*, 7, 271, 1974.
66. Reilly, D. and Burstein, J., *J. Biomechanics*, 8, 393, 1975.
67. Goldstein, S., *J. Biomechanics*, 20, 1055, 1987.
68. Kaplan, S., Hayes, W., Stone, J., and Beaupre, J., *J. Biomechanics*, 18, 723, 1985.
69. Townsend, P., Rose, R., and Radin, E., *J. Biomechanics*, 8, 199, 1975.
70. Ryan, S. and Williams, J., *J. Biomechanics*, 22, 1143, 1991.
71. Evans, G., Behiri, J., Currey, J., and Bonfield, W., *J. Mater. Sci.*, 38, 1990.
72. Hodgkinson, R., Currey, J., and Evans, G., *J. Orthop. Res.*, 7, 754, 1989.
73. Aro, H., Wippermann, B., Hodgson, S., Wahner, H., Lewallen, D., and Chao, E., *J. Bone Jt. Surg.*, 71A, 1020, 1989.

74. Markel, M., Wikenheiser, M., and Chao, E., *J. Orthop. Res.*, 8, 843, 1990.

75. Waanders, N., Senunas, L., Richards, M., Steen, H., Schaffler, M., Goldstein, S., and Goulet, J., *Transactions of the 41st Annual Meeting of the Orthopaedic Research Society*, 1995, 567.

76. Harten, R., Lee, F., Zimmerman, M., Hurowitz, E., Arakal, R., and Behrens, F., *J. Orthop Res.*, 15, 4, 1997.

77. Aro, H., Wipperman, B., Hodgson, S., Wahner, H., Lewallen, D., and Chao, E., *Transactions of the 34th Annual Meeting of the Orthopaedic Research Society*, 1988, 415.

78. Millett, P., Cohen, B., Allen, M., and Rushton, N., *Transactions of the 42nd Annual Meeting of the Orthopaedic Research Society*, 1996, 623.

79. Walsh, W., Asprinio, D., Sherman, P., Staebler, M., Trafton, P., and Ehrlich, M., *Transactions of the 19th Annual Meeting of the Society for Biomaterials*, Birmingham, AL, 1993, 123.

80. Bronk, J., Ilstrup, D., An, K.-N., Urabe, K., and Bolander, M.E., *Transactions of the 41st Annual Meeting of the Orthopaedic Research Society*, 1995, 587.

81. Harten, R., Zimmerman, M., Weiss, S., and Lee, F., *Transactions of the 21st Annual Meeting of the Society for Biomaterials*, 1995, 345.

82. Yasui, N., Kojimoto, H., Sasaki, K., Kitada, A., Shimizu, H., and Shimomura, Y., *Clin. Orthop. Rel. Res.*, 293, 55, 1993.

83. Harten, R., Zimmerman, M., Lee, F., and Alberta, F., *Transactions of the 42nd Annual Meeting of the Orthopaedic Research Society*, 1996.

84. Delloye, C., Delefortrie, G., Coutelier, L., and Vincent, A., *Clin. Orthop.*, 250, 34, 1990.

85. Frierson, M., Ibrahim, K., Boles, M., Bote, H., and Ganey, T., *Clin. Orthop.*, 301, 19, 1994.

86. Kojimoto, H., Yasui, N., Goto, T., Matsuda, S., and Shimomura, Y., *J. Bone Jt. Surg.*, 70B, 543, 1988.

87. Meunier, A., Riot, O., Kristal, P., and Katz, J.L., *Interface in Medicine and Mechanics*, Elsevier, London, 1991.

5

Techniques and Applications for Strain Measurements of Skeletal Muscle

Chris Van Ee
Duke University

Barry S. Myers
Duke University

5.1   Introduction

Determination of the constitutive properties and failure tolerances has been a principal activity of solid biomechanics for many years. Unlike engineering materials, however, the tissues and cellular structures of the body demonstrate uniquely challenging material behavior making the measurement of constitutive and failure properties difficult. In the face of these challenges a large experimental effort has occurred which has resulted in an increasingly accurate determination of the tensorial quantities upon which tissue properties and tolerances are dependent. In particular, the ability to measure stain in compliant, non-linear, hydrated, biologic tissues has become particularly refined. Yet, the selection of the appropriate formulation of strain, and the techniques used to measure strain vary widely. In this chapter, we present an overview of these techniques with particular reference to the measurement of strain in muscle.

5.2   Strain Theory

A wide variety of mathematical and experimental representations have been developed to quantify the components of strain within a deformable body. Most experimental methods quantity some component of motion and deformation and compute the strain components of interest based on a given strain formulation. In that regard, we present a review of deformation theory as given by Lai et al.[54] As a central underlying theme, strain measurement allows a quantification of the continuous, internal deformations of a material which are not the result of rigid body motions. Consider the point P_0 in the material located a distance $\mathbf{X}$ from the origin (Fig. 5.1). As a result of a change in position, $\mathbf{u}(\mathbf{x})$, the point P_0 is translated to a new position P at some subsequent time. The location of the point P is described by the vector $\mathbf{x}$ where $\mathbf{x} = \mathbf{X} + \mathbf{u}(\mathbf{X})$. In that regard, $\mathbf{u}(u, v, w)$ represents the motion of the point P_0 over time, where u, v, and w represent the components of motion in three orthogonal directions and are functions of position on the body, (x, y, z). To describe the state of strain at the point P_0 in the material we will

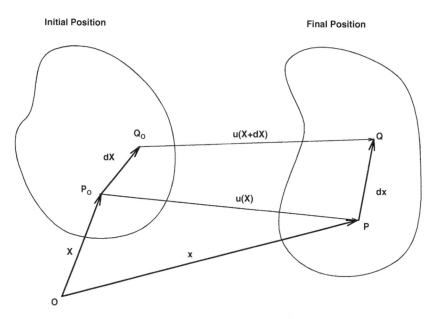

FIGURE 5.1   A body undergoes a transformation described by $\mathbf{u}(\mathbf{X})$. Internal deformation is quantified by examining the mapping $d\mathbf{X}$ and $d\mathbf{x}$.

examine changes in the length between P_0 and a closely neighboring point Q_0. The location of Q_0 is given by $\mathbf{X} + d\mathbf{X}$. At the same subsequent time, Q_0 translates to the new location Q. The new location of Q is given by $\mathbf{x} + d\mathbf{x}$, and its translation is described by $\mathbf{u}(\mathbf{X} + d\mathbf{X})$. These quantities are related vectorially by

$$\mathbf{x} + d\mathbf{x} = \mathbf{X} + d\mathbf{X} + \mathbf{u}(\mathbf{X} + d\mathbf{X}) \tag{5.1}$$

Recalling that $\mathbf{x} = \mathbf{X} + \mathbf{u}(\mathbf{X})$, this equation can be rewritten as

$$d\mathbf{x} = d\mathbf{X} + \mathbf{u}(\mathbf{X} + d\mathbf{X}) - \mathbf{u}(\mathbf{X}) = d\mathbf{X} + (\nabla \mathbf{u})d\mathbf{X} \tag{5.2}$$

where $\nabla \mathbf{u}$ is the displacement gradient. For Cartesian coordinates $\nabla \mathbf{u}$ is given by

$$\nabla \mathbf{u} = \begin{bmatrix} \partial u_1/\partial X_1 & \partial u_1/\partial X_2 & \partial u_1/\partial X_3 \\ \partial u_2/\partial X_1 & \partial u_2/\partial X_2 & \partial u_2/\partial X_3 \\ \partial u_3/\partial X_1 & \partial u_3/\partial X_2 & \partial u_3/\partial X_3 \end{bmatrix} \tag{5.3}$$

Further, we can define the deformation gradient, $\mathbf{F}$, as

$$d\mathbf{x} = \mathbf{F}\, d\mathbf{X} \tag{5.4}$$

Substituting this for $d\mathbf{x}$ in Eq. 5.2 and solving for $\mathbf{F}$ results in

$$\mathbf{F} = \mathbf{I} + \nabla \mathbf{u} \tag{5.5}$$

where $\mathbf{I}$ is the identify matrix. To quantify the deformation in a small neighborhood around $\mathbf{P_0}$ we compare the length of the vector $d\mathbf{X}$, denoted by the scalar quantity dS, to the length of the vector $d\mathbf{x}$ denoted ds, resulting in the following equation:

$$(ds)^2 = d\mathbf{x} \cdot d\mathbf{x} = \mathbf{F}\, d\mathbf{X} \cdot \mathbf{F}\, d\mathbf{X} = d\mathbf{X} \cdot \mathbf{F}^T\mathbf{F}\, d\mathbf{X} \tag{5.6}$$

Consider the special case of rigid body motion in which **F** is an orthogonal matrix comprised of orthonormal vectors such that $\mathbf{F}^T\mathbf{F} = \mathbf{I}$. Upon substitution into Eq. 5.6 results in $ds^2 = \mathbf{dX} \cdot \mathbf{dX} = dS^2$. Subject to this condition, the length is unchanged and the deformation gradient represents a rigid body rotation of the two points from their initial to their final positions. To define a strain tensor that is independent of rigid body motion, it is necessary to separate out the orthogonal portion of **F**. Using the polar decomposition theorem, we can write

$$\mathbf{F} = \mathbf{RU} = \mathbf{VR} \tag{5.7}$$

where **U** and **V** are positive definite symmetric tensors and **R** is an orthogonal tensor representing the rigid body motion. It can be shown that this is a unique decomposition in which **U** and **V** are known as the right and left stretch tensors, respectively. Using Eq. 5.76, it follows that

$$\mathbf{F}^T\mathbf{F} = (\mathbf{R}\ \mathbf{U})^T\ (\mathbf{R}\ \mathbf{U}) = \mathbf{U}^T\mathbf{R}^T\mathbf{R}\ \mathbf{U} \tag{5.8}$$

However, because **R** is orthogonal, $\mathbf{R}^T\mathbf{R} = \mathbf{I}$, and Eq. 5.8 simplifies to

$$\mathbf{F}^T\mathbf{F} = \mathbf{U}^T\mathbf{U} \tag{5.9}$$

Using this formulation, several strain tensors can be defined. The right Cauchy-Green deformation tensor, **C**, is defined as

$$\mathbf{C} = \mathbf{F}^T\mathbf{F} \tag{5.10}$$

The left Cauchy-Green deformation tensor, **B**, is defined as

$$\mathbf{B} = \mathbf{FF}^T \tag{5.11}$$

The Lagrangian finite strain tensor, **E**, is defined as

$$\mathbf{E} = 1/2(\mathbf{C} - \mathbf{I}) \tag{5.12}$$

Rewriting **E** in terms of the vector **u** results in the following expression:

$$\mathbf{E} = 1/2(\nabla\mathbf{u} + (\nabla\mathbf{u})^T) + 1/2(\nabla\mathbf{u})^T(\nabla\mathbf{u}) \tag{5.13}$$

Expressed in tensorial notation, this gives rise to the commonly used expression

$$E_{ij} = \frac{1}{2}\left(\frac{\partial u_i}{\partial X_j} + \frac{\partial u_j}{\partial X_i}\right) + \frac{1}{2}\frac{\partial u_m}{\partial X_i}\frac{\partial u_m}{\partial X_j} \tag{5.14}$$

where m is a summed index. Each of these quantities results in a second order tensorial definition of strain tensor. Each component of the strain tensor can be determined if the full three-dimensional displacement field, $\mathbf{u}(u, v, w)$, is known. Fortunately, in many situations, it may not be necessary to use such a generalized representation of strain. However, it should also be recognized that, even in problems in which only the uniaxial normal strain is to be determined, the three-dimensional displacement field may require characterization. Examination of the Lagrangian representation of strain for the characterization of the axial strain, $E_{xx}(x, t)$, illustrates this phenomenon:

$$E_{xx}(x, t) = \frac{\partial u}{\partial x} + \frac{1}{2}\left[\left(\frac{\partial u}{\partial x}\right)^2 + \left(\frac{\partial v}{\partial x}\right)^2 + \left(\frac{\partial w}{\partial x}\right)^2\right] \qquad (5.15)$$

In this situation, gradients of the all three deformation terms are required to characterize a single normal strain. However, if the spatial gradients of the displacements are small, the so-called small strain problem, the higher order terms may be neglected and the commonly used one-dimensional small strain formulation

$$E_{xx}(x, t) = \frac{\partial u}{\partial x} \qquad (5.16)$$

results. This formulation is only valid if the higher order terms remain small with respect to the first order gradient terms. For example, a displacement gradient of $\frac{\partial u}{\partial x} = 0.15$ results in a Lagrangian strain of 0.161. Using small strain theory, an underestimation of strain of 0.011 results, representing an error in strain measurement of 7%.

Another commonly used convention is the stretch ratio. The stretch ratio, λ, is defined as

$$\lambda = \frac{\Delta L}{L} \qquad (5.17)$$

The stretch ratio is not a tensorial component of strain. However, the stretch ratio is a convenient way to report finite uniaxial deformations with a more obvious direct physical significance. Further, if measured in the appropriate directions, three stretch ratios can completely define other tensorial definitions of strain, like the left Cauchy-Green strain tensor. This approach has been used together with the assumption of incompressibility to model rubber, mesentery, muscle, and brain.[18,21,52,70,84,107,110]

Ultimately, the choice of the method used to define strain will depend on the type of problem, the constitutive formulations, and the available experimental data. Regardless of the strain formulation, however, careful definition of the strain components measured in a given study is a requirement for the generation of meaningful data.

5.3   Experimental Considerations

In making mechanical measurements of soft tissue, a need to reproduce the *in vivo* environment is required if meaningful results are to be obtained.[14,30,44,80,116] For example, the constitutive behavior of many biological tissues has been shown to be both temperature and hydration sensitive.[30,80] Mechanical stabilization, or preconditioning, a process in which the tissue is cycled repeatedly at the onset of a test battery, has also been shown to influence soft tissue behavior, and is necessary for generation of repeatable constitutive data.[36,67,116] Soft biologic tissues are also sensitive to the loading rate and duration which can vary from a few milliseconds during impact injury to seconds and hours in activities of daily living.[47,66,79] Several tissues, most notably skeletal muscles, undergo very large strains during physiologic loading. Zajac reports that the tibialis anterior of the rabbit experiences a change in length between 15 and 20% during hopping.[120] Changes in length between 10 and 50% have been reported in the lower extremity of the cat during gait.[34] Using a computational model, Merrill et al. suggested that the muscles of the human cervical spine elongate as much as 40%.[71] Additionally, as experimental test specimens are often removed from their anatomic origins and insertions, specific attention to the effects of the *in vitro* conditions is required. That is, the loads applied by the experimental apparatus must mimic those applied in the *in vivo* environment, and the strain measurement technique must account for the effects of the nonphysiologic end conditions on the measured deformation. A novel solution to this problem, which allows for the appropriate selection of end loads, and provides estimates of *in vivo* loads is the measurement of strain *in vivo*. Hawkins et al. measured strains of rabbit medial collateral ligament *in vivo* where force measurements are impossible because of the redundant construction of the knee ligaments. The ligament was then tested *in vitro* and the loads required to impose similar strains were determined and used as estimates of the *in vivo* loads.[44]

Accurate measurement of soft tissue strain is made more complex by the material heterogeneity of biologic structures in which the soft tissues of the body often show strong regional variations. In muscle, this includes the anatomically distinct regions of the tendons, aponeuroses, and the muscle fibers. For example, testing of whole muscle tendon units reported in Lieber et al. resulted in average normal strains in the mid-tendon, tendon-bone interface, and apnoneurosis of 2, 3.4, and 8%, respectively.[60] Variations in the time-dependent responses of the components of muscle add a rate sensitivity to the regional variations in strain. According to Best et al., differences in mid-muscle belly Lagrangian strains of 30% are reported as the rate of elongation of the tendon increased from 40 to 100 cm/sec.[8] Strain fields also vary from regional variations in material properties within a given muscle. As a result, even small test specimens can have wide variations in their constitutive properties resulting from an inhomogeneous strain field, despite having uniform test specimen geometry. Mechanical measurement in muscle is also made more complex by the variations in area, and therefore stress, along the length of the test specimen, as occurs commonly during whole skeletal testing.[77]

Measurement of skeletal muscle strain is made uniquely complex by the tissue's need to have proper reproduction of its *in vivo* environment. Specifically, the mechanical properties and hence deformations of skeletal muscle are dependent upon both a passive extracellular matrix, and a dynamic, metabolically dependent, intracellular protein interaction. In contrast, many passive biological tissues such as bone, ligament, tendon, and skin derive their mechanical properties exclusively from extracellular matrix interactions. The intracellular responses depend strongly on the normal function of the cell, in addition to the correct electrical excitation of the membrane. For these reasons, both nutrients and electrical potentials must be supplied to the muscle during testing through either an intact neurovascular supply or through a nutrient-rich bath using specimens of sufficiently small size to allow for the delivery of nutrients by diffusion along. Gaining access to the tissue for deformation measurement is made increasingly difficult by these additional experimental design considerations.

Considering only the passive responses, skeletal muscle still represents a complex experimental challenge. Several investigators have shown that the passive mechanical properties of skeletal muscle are a result of mechanical load carried by both intracellular and extracellular proteins.[65] As a result of postmortem intracellular proteolysis,[102] the intracellular load carrying potential is lost, and the ability to study cadaveric muscle is profoundly limited. Indeed, skeletal muscle experiences large changes in its mechanical properties postmortem despite using currently accepted methods for storage of bone, ligament, tendon, cartilage, and skin.[26,27,35,58] These changes include a decrease in stiffness of 47%,[35] and a decrease in strain at failure of 44%.[58] Fitzgerald[26,27] reported the onset of changes in skeletal muscle mechanical properties within the first 6 hours of death. Our own laboratory studies characterizing the mechanical properties of skeletal muscle in the rabbit tibialis anterior support these observations (Fig. 5.2). The onset of rigor can be seen as early as 6 hours postmortem. By 8 hours postmortem, stiffness increased 600% over the live passive muscle stiffness. Following this initial stiffness increase, which peaks at 12 hours postmortem, a gradual decrease in stiffness occurs over postmortem hours 12 to 62.5 resulting in a final muscle stiffness which is 28% less than the live passive muscle stiffness.

Despite using methods that have been developed for storage of bone, ligament, and tendon, there are no standardized methods to store whole muscles or cadavers through the postmortem period that maintain the integrity of the microstructure and associated mechanical properties under the action of postmortem enzyme activity. For this reason, investigators have developed methods to store and test single muscle fibers.[104,105] The fibers are mechanically[78,86,87] or more often, chemically skinned[23] disrupting the sarcolemma while leaving the myofibrillar structure intact. The fibers are then placed in a storage solution where the enzyme activity is modulated, resulting in the preservation of the intracellular protein structure. The fibers' mechanical properties are thereby stabilized, and the tissue can be frozen for months and tested without alterations in the mechanical properties. Further, the disruption in cell membrane

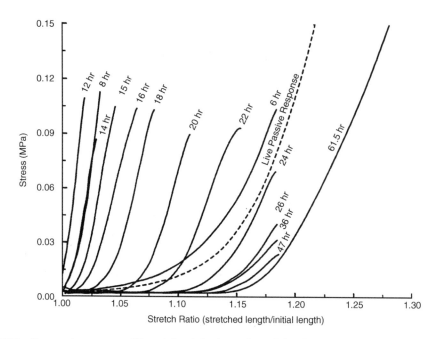

FIGURE 5.2 Structural responses of live passive skeletal muscle, and the changes in passive properties that occur over time. The onset of rigor can be seen as early as 6 hours postmortem. By 8 hours postmortem, stiffness has increased 600% over the live passive muscle stiffness. Following this initial increase, which peaks at 12 hours postmortem, a gradual decrease in stiffness occurs over postmortem hours 12 to 61.5 resulting in a final muscle stiffness which is 28% less than the live passive muscle stiffness.

integrity allows for more direct access to the interior of the cell, allowing cell activation, relaxation, and rigor by altering the fluid bath which surrounds the cell.[75]

Maintaining tissue hydration for *in vivo* experimentation is required during testing to avoid changes in mechanical properties due to drying.[30] This has been accomplished experimentally using repeated applications of normal saline,[8] continuous normal saline drip,[43,58] application of a thin film of paraffin oil,[108] immersion within a fluid bath,[60] covering with moist gauze,[10] or enclosing within a 100% humid environment. Within muscle, an intact neurovascular supply aids in tissue hydration; however, care should be taken to keep the surface moist. Results of hydration methods for ligament and tendon are conflicting. Woo et al. report dehydration of tendon to cause increased stiffness and increased failure stress.[116] Consistent with this finding is a study by Haut and Little where immersed ligaments were found to have a decreased modulus when compared to ligaments only slightly moistened.[40] In contrast, a later study by Haut and Powlinson[42] reported increases of 43 and 61% of tensile strength and stiffness for immersed specimens compared to specimens moistened by a drip method of the same solution. A source for some of these inconsistencies may be the effect of strain rate. In a study by Haut and Haut, changes in water content of tendon had no effect on mechanical properties at a strain rate of 0.5%/s, but at a strain rate of 50%/s, tissues with higher water content were 20% stiffer.[41] In an effort to better understand the *in vivo* conditions and the effects of different immersion solutions, Chimich et al. measured water content of ligament immediately after sacrifice and compared it to ligament water content after soaking in phosphate buffered saline (PBS) and 2, 10, and 25% sucrose solutions.[14]

Water content immediately after sacrifice was found to be 64% ± 6% while contents of PBS and the 2, 10, and 15% solutions were found to be 74, 69, 60, and 51%, respectively. Relaxation tests were performed upon the four immersed tissues and higher water contents were associated with a greater degree of load relaxation. However, tests on the properties of the fresh ligament were not reported. The need to reproduce the *in vivo* hydration within articular cartilage is also documented. Elmore et al. demonstrated the *in vivo* recoverability of articular cartilage if immersed in a balanced salt solution.[24] In contrast to tendon and ligament, the recoverability of articular cartilage was closely linked to the tonicity of the solution. Previous experimenters testing in air had reported residual deformations after the load had been removed and had termed this experimental artifact the "imperfect" elasticity of articular cartilage. Appropriate selection of the hydration in the extra-specimen testing environment and careful consideration of the importance and physiologic relevance of tissue swelling and ion movement are therefore equally important requisites to the measurement of meaningful *in vitro* strains.

Temperature has also been shown to significantly affect the properties of skeletal muscle.[80] While the properties of collagen have been shown to be relatively insensitive to changes in temperature ranging from 0 to 37°C,[3,90] the properties of muscle have been shown to change over a range in temperature from 25 to 40°C.[80] Specifically, stiffness decreased 11 to 15% while deformation to failure was unchanged. To standardize conditions of testing environment, a majority of investigators have chosen to use environmental chambers to control both humidity and temperature.

The ability to determine the properties of soft tissue is limited because the loading conditions along the tissue boundary are often unknown or difficult to recreate. Yet when the tissue is excised and evaluated *in vitro*, recreation of these *in vivo* loads and boundary conditions is requisite to the generation of meaningful constitutive data. Indeed, several investigations have shown that measured constitutive and tolerance data depend upon the methods by which the load is applied.[10,39,116] To realize appropriate loading conditions *in vitro*, investigators have developed a wide variety of gripping devices and methods.[11] Directly gripping the specimen is among the most common methods used in biomechanics. Most grips compress the tissue in a clamp with the hope of distributing the load such that the specimen neither slips in the clamp nor suffers excessive damage in the clamp. Clamp designs are numerous and include direct clamping by smooth or patterned metal grips, or sinusoidal shaped grips.[10] Capstan grips, in which the tissue is wrapped around a cylindrical shaft and clamped, are also employed. These have the advantage of decreasing the load on the tissue at the clamp at the expense of allowing slip in the specimen around the capstan. However, assuming a constant coefficient of friction between capstan and grip can provide estimates of the tissues spoolout during loading. Other approaches include the use of cyanoacrylate adhesive, embedding the tissue in polymethacrylate, and freezing the tissue directly to the clamps.[88,95] To minimize gripping effects, investigators have tested bone-tissue-bone specimens to evaluate the mechanical properties of ligament, tendon, and muscle.[11,13,35] Gripping of bone results in more physiologic load distribution in the soft tissue as the tissue's anatomic origin and insertion are preserved. Further, the bone is more easily gripped without risk for slip or mechanical failure at the grip site.

Variations in strain distribution near the clamp are also thought to influence results. Saint-Venant's principle states that end conditions whose resultant force and couple are zero will not influence the state of stress and strain at distances that are large compared to the dimension over which the load is applied.[29] Experimentally, this implies that any state of stress with an equivalent resultant force and couple at the grips will not alter the expected state of stress or strain remote from the clamps. Thus, for well-behaved test specimens, the midsubstance stress and strain distribution will not be affected by the end conditions and will approach the theoretical solution remote from the point of application of the load as though an ideal load distribution had been applied. Experimentally, this requires long, slender specimens and a path for load redistribution within the specimen that cannot always be realized with biomechanical specimens. The significance of gripping effects was noted by Butler et al. Midsubstance tendon strains were found to be 25 to 30% of strains near the grips or at the bone-tendon junction.[10] Haut reported changes in maximum failure strain of small collagen test specimens as a function of specimen length.[39] Surprisingly, Haut reported maximum strains at failure of 8.1% ± 0.3%, 8.5% ± 0.6%, and 17.6% ± 1.2% for specimen gauge lengths of 100, 50, and 10 mm, respectively. Interestingly, the failure load was insensitive to specimen length.

Increasing interest in the biomechanical behavior of cells has resulted in development of methods to grip and measure strain on the surface of cells. Barbee et al. measured strains on the ventral cell surface of cultured vascular smooth muscle cells.[7] Cell deformations were imposed by deforming the compliant polyurethane substrate to which the dorsal surface of the cell was adherent. As in testing of larger structures, this adhesion gripping technique has been criticized. Concerns over failure of the cell to adhere to the membrane or cell injury as a result of membrane deformation are commonly raised. However, Anderson et al. provide data on the strains of cultured lung, muscle, and bone cells which refute these assertions.[2] They report a close agreement between cell strain and estimated membrane strain. They also found no changes in membrane permeability, a sign of cell damage with strain, as measured by both fluorescent and trypan blue straining techniques.

Realization of optimal clamping technique without either slip or inappropriate failure with the clamps remains a challenge in biomechanical testing. With regard to strain measurement, the use of more complex experimental measures of strain than simple grip-to-grip excursions can often decrease the demands placed on clamp design and performance. That is, by measurement of full-field strain, the effects of the end conditions on deformation can be accounted for and less constraining clamps or enlarged grip surfaces can be employed. This is particularly relevant in failure testing in which the issue must fail at sites remote from the clamp in order to be considered meaningful. Thus, the ease with which the specimen can be coupled to the test apparatus often serves to define the type of strain measurement technique which should be used.

Most measures of strain require, or are based on, a definition of the undeformed geometry or a reference length in uniaxial problems. While often easily and uniquely defined for engineering materials as the specimen's zero load state, such definitions are less apparent in biologic tissues. Soft biologic tissues are typically compliant, nonlinear, strain-stiffening materials which show a large low-load region at small strains. As a result, small differences in the tare load (preload) used to mount the specimen can result in profoundly different initial positions.[103] Further, creep effects under constant load can result in shifts in a load-based initial reference length over the duration of testing. In other words, subject to a load, the reference length changes with time. In order to manage this problem, authors have suggested fitting the load-elongation data and extrapolating to a no-load length.[89] Other techniques to help define a reliable reference length in muscle include using a gross anatomic position, a microanatomic position, or a physiologic position. Garrett et al. established the initial length of the rabbit tibialis anterior based on an anatomical position by measuring the muscle length when the knee and ankle were both flexed to a joint angle of 90°.[31] *In situ* sarcomere length may also be used.[62] The force-length relationship of stimulated skeletal muscle has also been used to define a physiologic initial position. Specifically, by electrically twitching a muscle at various lengths the initial position of the muscle can be defined as the length at which maximum twitch force is measured.[106] In the same context, the minimum length and maximum length at which no force is generated in response to an electrical stimulus may also be used as a reference length. Relationships between sarcomere length, whole muscle length, and twitch properties may allow for comparisons of data generated using these different gauge lengths.[16,120] Measurements of the reference state of cardiac muscle are often based on gating the position-time histories with the cardiac cycle or cardiac electrical stimulation.[38,111,112] In either case, clear definition of a reproducible reference length is requisite in strain measurement.

5.4   Displacement Measurement

Selection of the displacement measurement strategy is predicated on several factors including the physical constraints imposed as a result of the need to create a physiologically appropriate testing environment and the formulation of strain to be reported. Design considerations include the degree of intrusion imposed by the measurement system on the tissue, the required measurement accuracy, the frequency content of the experimental data, and the frequency response of the measurement system. Also of importance are the need to measure surface or internal deformations and the need for real-time displacements or post-test data analysis to determine displacements. Other considerations include the cost of the system, the ease of use, the number of dimensions to be measured, and the need for average or full-field measures of displacement. In light of the diversity of design considerations, it is of little surprise that the techniques employed vary widely.

Like many other tissues in the body, skeletal muscle exhibits a hierarchial structural organization. Among the most significant functional building blocks of whole skeletal muscle is the sarcomere. Indeed, whole muscle structural properties may be derived directly from the organization of sarcomeres. Specifically, force production is proportional to the numbers of sarcomeres arranged in parallel, while the maximum tendon velocity is proportional to the number of sarcomeres in series. Sarcomere lengths range from 1.5 to 4.0 µm and both stimulated and passive muscle responses have been shown to be functions of the sarcomere length.[19,65,74,114] In that regard, measurement of constitutive properties as a function of sarcomere length as opposed to strain is a common practice.[65,74] The two most common methods to measure sarcomere length are laser diffraction and microscopy.

Microscopy affords direct observation of the tissue and has been useful in determining complex microstructural interactions.[20,33] However, the size and proximity of the optics often preclude sarcomere measurements during mechanical testing except for single muscle fiber preparations. In an effort to automate sarcomere length measurements, De Clerk et al. captured microscopic images of muscle cell striations by video.[20] Sarcomere lengths over a population of cardiac cells were determined by using a Fourier transform in which the spatial frequency of the spatially periodic sarcomeres was determined. The laser diffraction technique also makes use of the periodic structures of the sarcomeres in the muscle fiber. Acting as a diffraction grating, a monochromatic light passing through the fibers produces a diffraction pattern in which the distance between the first set of parallel lines is proportional to the sarcomete length.[61,62] The laser affords the advantage of being able to measure average sarcomere lengths through a thicker sample than compared to microscopy. As a result, dissection times are substantially reduced using laser techniques. Laser techniques have the additional advantage of adaptability for use in measuring *in situ* sarcomere length. Fleeter et al. used laser techniques to measure sarcomere length of human forearm muscles *in situ* for use in tendon transfer surgeries to determine optimal muscle length before attachment of the tendon.[28] Trestic and Lieber were also able to make *in situ* sarcomere length measurements. Using the frog gastrocnemius muscle, in which the central tendon typically precludes whole muscle diffraction measurements, these authors showed that diffraction measurements on partially dissected bundles containing approximately 100 fibers compared favorably with those of the intact muscle.[106] Lieber et al. were also able to automate the laser diffraction method to measure sarcomere length in single muscle fiber tests. They report a frequency response of 3.8 kHz and a measurement accuracy of 0.043 µm.[61] Thus, in single fiber measurement, laser diffraction techniques provide higher frequency response but somewhat lower spatial resolution than microscopy techniques.

Analogous techniques have also been used on tendons. Sasaki and Odajima measured microstructural deformation in tendons using X-ray diffraction.[91,92] Using a wide-angle X-ray diffraction technique, reflections were produced corresponding to distances between neighboring amino acids along the helix of the collagen molecule. Using small-angle X-ray diffraction, the fibrillar organization of whole collagen molecules was measured.[13] Interestingly, these authors have also used microscopy techniques to measure tendon strains, and have reported an accuracy on the order of ±0.1%.[91] These measurements have led to the development of structural models based on true microstructural interactions of collagenous tissues.[17,56,57]

In contrast to microstructural measurements, tissue deformation can be measured on a macroscopic scale using transducers which directly measure displacement. Grip-to-grip measures of displacement are among the easiest and most commonly used. This method typically measures actuator motion for a specimen with a grip mounted to the actuator, and another grip mounted to a fixed platen. The actuator is usually instrumented with a linearly variable differential transformer (LVDT) which serves as a feedback signal for the control algorithm. This signal is split, acquired digitally, and used as a measure of tissue displacement. In that regard, the tissue motion can be measured without the use of additional instrumentation. While a wide variety of transducers can be used to measure displacement, the LVDT is the most common, owing to its excellent frequency response, variable sensitivity, stability, and fatigue-resistant design. The devices are AC transformers which detect changes in the magnetic field as a magnetically permeable core is moved within the field. As a result, the moving core is not in direct contact with the housing, and fatigue effects are minimized. The accuracy of the displacement measurement of a typical LVDT is between 0.5 and 0.25% of full scale stroke. Full scale stroke ranges from as little as 0.1 to 650 mm. Data can be acquired at frequencies up to 10% of the excitation frequency and excitation frequency varies from 60 Hz to 25 kHz, although higher frequencies are attainable.[25] Linear potentiometers can also measure displacement; they are considerably less expensive, approximately 10% of the cost of an LVDT. However, they lack the sensitivity and flexibility of an LVDT and are prone to fatigue as they operate through a contact mechanism. For strictly static measurements, dial gauges have also been used but viscoelastic effects can confound static measurement techniques.[11] Dial gauges typically provide measurement accuracies on the order of ±0.0125 mm.

While often used because of the ease of testing a whole muscle tendon unit, grip-to-grip measures are not without limitations. The method is predicated on the use of a stiff load frame, such that test frame deformation does not corrupt the displacement measurement. For example, Lieber et al. reported a system compliance of 1.3 µm/g when measuring displacements of 13 mm under peak loads of 2000 g.[59] Another limitation with this method is the heterogeneity of the tissue between the grips. In testing whole muscles with long tendons, the deformation of the tendon and aponeurosis can be significant, despite subjection to smaller strains than the muscle. Trestik and Lieber report the results of an experiment in which the gastrocnemius-Achilles tendon complex was stretched.[106] Strain within the complex varied from 2% in the tendon up to 8% in the aponeurosis when muscle strains of 15% were observed. Clearly, an assumption that the measured displacement could be assigned directly to the muscle would result in substantive errors in assumed muscle strain.

Of greatest concern with this technique is the possibility of slip of the specimen within the grips, particularly during failure testing. In general, larger measurements of deformation, and therefore strain, are recorded when using bone-to-bone to grip-to-grip methods rather than when displacements are measured within the tissue midsubstance.[10] Studies by Haut, Zernicke et al. and Butler et al. found significant differences in the calculated modulus, failure strain, and failure strain energy density when using grip-to-grip methods as compared to regional measures of displacement.[10,39,121] Even when slip is minimized through the use of good gripping technique, the effects of gripping on local tissue deformation can be significant. Zernicke et al. reported strains as great as 60% close to the grips while measured strains remote from the grips were less than 15%. After testing gracilis, semitendinosus, fascia lata, and patellar tendon-bone units they reported that grip-to-grip measurements of strain were 3.2 times larger than averaged regional values through the midsection of the specimen.[121]

The most common solution to the errors associated with grip-to-grip measurement in engineering has been the use of dumbbell-shaped specimens with a midsubstance over which displacements are measured. Clip gauges and extensometers have been commonly used to measure midsubstance displacements, though they require more robust specimens than are often found in biomechanics. Clip gauges have been developed using spring steel and other materials.[11,13] They make use of a strain gauge mounted on an arched flexible strip formed to the desired gauge dimensions. Clips at the ends of the strip are coupled to the tissue using glue, sutures, or clamps. Design and selection of these devices ultimately become a trade-off between intrusion (the loads imposed on the system by the measuring device which can alter the deformation and damage the tissue) and displacement sensitivity and frequency response. Potentiometers and LVDTs can also be used in similar applications, although the need to maintain the coaxial orientation of the core and housing can prove difficult.

Because of its versatility in design and very low cost, the foil strain gauge has also enjoyed popularity as a direct contact measurement of tissue strain, although its use is limited to higher modulus materials like bone.[73] Strain gauge concepts have been adapted for use with more compliant soft tissues like tendon and ligament. For example, liquid metal strain gauges make use of mercury-filled silastic tubes that are sutured directly to the surface of the tissue. Described in detail by Brown et al. and Meghan et al., these devices operate through the same mechanism as the foil gauge.[9,68] As the tube's length is changed, the cross section is altered and the resistance changes. Meglan et al. used this gauge to measure anterior cruciate ligament strain.[69] Van Weeren et al. measured the *in vivo* strains of the peroneous tertius tendon of a horse.[109] Brown et al. measured the dynamic performance of these gauges and their applicability to measurements of soft tissue strain.[9] The gauges were found to be linear up to 40% change in length. In addition, gauges exhibited frequency independent response up to 50 Hz for a cyclic 20% change in length. Stiffness of the gauge was found to be 0.024 N/mm, which is approximately 4% that of tendon. However, the shelf life of the gauges is limited to about 3 months due to the oxidation of the mercury through the silastic tubing.

Arms et al. used a Hall effect sensor to estimate strains in the medial collateral ligament.[6] A semiconductor device measured the proximity of a permanent magnet attached to the tissue. Cholewicki et al. also employed a Hall effect device to measure motion between the facet surfaces of the intervertebral joint.[15] The device's range of motion was 4 to 12 mm with a reported accuracy of 0.025 mm. Frequency response of the sensor was reported to be 20 kHz; however, mass effects associated with the guide track, sensor, and magnet will likely limit frequency performance to a value considerably below the sensor response. Additionally, the calibration of this device was nonlinear, making its use somewhat more difficult. While the cost of a Hall effect sensor is typically less than a dollar, the mounts must be designed and assembled by the investigator. Commercially available sensors are available; the cost is approximately $900.

Villarrel et al. and Omens et al. describe the use of three piezoelectric crystals to determine planar displacements of the left ventricle of the heart. Each crystal both receives and transmits signals to the other two crystals resulting in measures of length based on assumptions regarding transmission velocity in the media. Frequency response of the system was 375 Hz. Accuracy of the system was not calculated, but results were similar to those derived from using biplane coneradiography.[82,111]

George and Bogen report the design, construction, and use of a novel biaxial fiberoptic strain gauge system.[32] The system employs 0.76 mm diameter fiber optic cables which are inserted transversely through the substance of the tissue and direct light onto a large silicon photo diode which tracks the point at which the light contacts the diode array. The system is synchronized so that as each fiber is illuminated in order, the diode output is sent to a multiplexer resulting in a system frequency response of approximately 3 kHz. The authors noted that hardware cost was approximately $1000. That figure did not include the circuit design, construction, testing, or calibration. The authors used this prototype device to measure tissue strains as large as 40% during a biaxial test of a flat section of the ovine right ventricle. As the fibers are transversely mounted through the section of tissue, their intrusion on the deformation was thought to be minimal. The accuracy and calibration of this method are not reported. Further, the errors due to optic fiber rotation or distortion across the cross section of the tissue that would result from a nonuniform strain field were not discussed.

Because direct contact methods using transducers are invasive, and rarely provide full-field measures of strain, noncontact methods have become increasingly more popular. These methods typically track the position of tissue markers over time to determine displacements at discrete points of the tissue and are commonly used to determine the full-field strain variation across the tissue. Obtaining tissue marker contrast and quantifying marker position are achieved using a wide variety of tools and techniques including clinical imaging systems and optical methods. Clinical imaging systems have the advantage of being able to capture images of the tissue when the region of interest cannot be directly visualized using optical methods.

Biplanar cineradiography is among the most common and oldest techniques, and has been used in a variety of applications including determining the deformations in a three-dimensional space of the beating canine heart *in vivo*.[38,112] A number of studies have investigated the methods used for deriving the three-dimensional space from the planar images and the results of a parametric analysis of the errors associated with biplanar cineradiography have been reported.[1,45,63,72,98,115] Hashima et al. used this method to track the positions of an array of 25 lead beads (1.0 mm in diameter) sewn to the epicardium of the left ventricle in an array approximately 5 to 10 mm apart.[38] Frames were captured at a rate of 120 Hz but other studies report using frequencies up to 3000 Hz.[37] To calibrate the system, several 1.0 cm long radiopaque rods were placed within the field and imaged. In a similar study, Waldman et al. found the error in the reconstruction of the three-dimensional displacements using biplanar cineradiography to be 0.3 mm and was limited primarily by digitization errors.[112] Pin cushion distortion (warping along the edges of the image) and the cone effect (magnification dependent on distances from the focal plane) were relatively small compared to the digitizing errors (0.05 and 0.1 mm, respectively). In a later study examining errors associated with this method, Waldman and McCulloch reported that marker positions could be located in three-dimensional space with a standard deviation of 2.5% of the full-field of view using biplane cineradiography.[113]

The recent use of magnetic resonance tagged images has eliminated the need for the invasive implantation of radiopaque markers into the tissue associated with biplanar cineradiography. The technique is based on locally perturbing the magnetization of the myocardium with selective radio-frequency saturation, resulting in multiple, thin tag planes. The imaging plane is orthogonal to the tag planes and the intersections of these planes result in dark stripes whithich deform with the tissue. The intersections are tracked, resulting in a spatial history of the tissue deformation. Young et al. reported using this method to track the deformations of the human heart. A total of 3100 points were tracked with an RMS error of half a pixel or 0.47 mm.[119] Studies by other investigators reported similar point location accuracy of approximately 0.3 mm.[53,81] An obvious advantage of the magnetic resonance tagging methods is the noninvasive ability to track the motion of an entire tissue volume including the tissue substance and the tissue surface *in vivo*. Current techniques, however, remain limited by the frequency and duration over which these images may be obtained and the number of institutions equipped to perform these measurements.

Optical methods have been widely used and include still photography,[100] video,[43,36,85,121] and CCD (charge coupled device) cameras.[8,108] These systems track optical markers or tissue landmarks on the surface of the tissue to determine displacements. A great number of different optical markers have been used. Markers vary in size, shape, color, and material. Selection of an appropriate marker aids in tracking, improving accuracy, and minimizing the effect of the marker on strain profile. Accuracy of each of these techniques seems most tightly coupled to reliably determining an exact marker location during digitization.[100]

Improving marker contrast to gain accuracy and ease of marker tracking has been studied extensively. Non-reflective markers can be made by blackening a surface with sulfide, ink, or paint. Fluorescent markers have also been used.[7,99] Smutz et al. used fluorescent markers illuminated by an ultraviolet light source in an otherwise dark room. Others have relied on tissue-mounted LEDs. These methods have the advantage of allowing band pass filtering at the excitation frequency to improve contrast between the markers and the background.[46,108]

In addition to marker contrast, the attachment of the marker to the tissue requires careful consideration. Hoffman and Grigg used stopcock grease or mineral oil to attach 600 μm disks to the posterior joint capsule of the cat knee.[46] Other methods to attach markers are the use of histoacryl glue,[85,108] cyanoacrylate glue,[100] and small sutures.[12,38] The attachment of external markers to the tissue is not without consequences. Barbee et al. validated their method of bead attachment to single smooth muscle cells by microscopically examining the substructure with and without the beads attached to insure that the cell did not reorganize with the addition of the 10 μm diameter microspheres.[7] Investigators have also stained the tissue of interest directly with paints or ink to avoid the problems of external marker attachment. Elastin stain,[106] Verhoeff's stain,[22] and India ink have all been used.[8,10,121] While decreasing intrusion, stains typically create irregular marks with varying signal intensities whose locations may be more difficult to determine.

A majority of these methods involve placing the marker on the surface of the tissue, thus only providing data about the displacements at the surface. In an effort to measure the deformations within the substance of articular cartilage, Schinagl et al. used fluorescently stained cell nuclei as markers.[93] Nuclei at different depths were then tracked using transmitted and epifluorescence microscopy.

In order to track the marker displacement, an image capture technique needs to be implemented. For static testing, a low method to track marker displacement is photography.[100] The shutter time or acquisition time of a single frame must be quick enough to avoid blurring of the markers in the image. Typical video systems have a frequency response of 30 Hz, but split frame video at 60 Hz is also common.[118] To avoid blurring of moving markers during the long exposure times of conventional video. Prinzen et al. incorporated the use of a xenon strobe which was triggered by a video frame pulse.[85] Hoffman and Grigg described a method using a high-sensitivity television camera, a trinocular microscope, and a video image frame grabber to store the digitized image on a microcomputer for post-processing.[46] Other investigators have recorded their images using CCD cameras that allow image acquisition rates up to and greater than 10 kHz.[8,108] Microscopes can be used in conjunction with video or CCD cameras to track the positions of small particles.[7,93] Polaroid filters have also been used to eliminate unwanted glare associated with moist tissues.[85]

For three-dimensional measurements of displacement, two or more image views are obtained and reconstructed through the use of direct linear transformation methods.[6,1,50,51,64,97] Sirkis and Lim described the equations used for a direct linear transform assuming a pin-hole camera, and investigated the role of possible errors in the process.[97] Luo et al. examined the effects of changing the angle between the two cameras used to quantify the three-dimensional space.[64] A lower limit of the pan angle was found to be 20 degrees. No improvement in accuracy was noted as pan angles increased to 40 degrees and testing at larger angles was not reported. As often occurs in biomechanical systems, planar motion data are desired from a curved or slightly irregular surface. Waldman and McCulloch and others investigated errors due to single plane vs. multiplane imaging of the curved surfaces of the heart, and provide guidelines on the maximum allowable curvature for single imager system.[82,113]

Identification and tracking of markers from captured image data have received considerable attention as it can be both time intensive and error prone. Automated edge detection, grid tracking algorithms, and image correlation techniques have all been refined to improve the speed and accuracy of this time-intensive process.[22,55,85,94,99,106,117] One of the first automated methods of optical strain measurement was the Video Dimension Analyzer (VDA).[55,106,117] Horizontal lines stained on the tissue are captured by a video camera and displayed on a monitor with an electronic dimension analyzer which outputs a voltage based on the distance between the two lines. The frequency response of this system is approximately 20 Hz and the results are displayed in real time. The drawbacks of VDA are that only strains in one dimension are measured and the strain is averaged between the two markers. Lam et al. reported calibration of the VDA. Four different experiments were conducted to measure the accuracy of the method. First, the effect of changing camera and object distance was measured. The second and third tests measured the influence of imaging through the wall and saline environment of a test tank and the effect of changing the angle of incidence. The final test was to measure the dynamic response of the system. The accuracy of the tracking device at locating the edges of the marker lines was found to dominate the error analysis. Variations due to the above perturbations of the system did not significantly affect accuracy and overall the VDA was found to be accurate to 1% strain.[55]

Derwin et al. described an automated method to determine uniaxial strain in which horizontal stain lines spanning the cross section of the tissue are tracked through vertical displacements. To avoid discontinuities or breaks within the line, the image was smoothed by convolving the image intensity with a Gaussian function. Next, a gradient was calculated in the direction of displacement (direction must be given by the user). The gradient was then thresholded to give areas of positive and negative slope corresponding to each edge of the line. The edges were then averaged and tracked through sequential images resulting in a displacement history.[22] Prinzen et al. described a method by which 43 paper markers on the surface of the heart were automatically sorted and tracked in a single plane. Strain distribution is then determined by separating the region into triangles and computing the planar strain components from the changes in the lengths of the sides of each triangle.[85]

To improve marker recognition during digitization, images are often smoothed, sharpened, or enhanced. Smoothing reduces the signal intensity variation between nearby pixels, and is often used to reduce noise.[49] This has the effect of eliminating pixel values that are unrepresentative of their surroundings. Median filtering is a local smoothing process in which a pixel's intensity is replaced with the median of neighboring pixels. Since the median value must actually be the value of one of the pixels in the neighborhood, the median filter does not create unrealistic pixel values when the filter straddles an edge. For this reason the median filter is much better at preserving sharp edges than the mean filter. It is particularly useful if the characteristic to be maintained is edge sharpness.[4] Image sharpening to better define the edges of the markers is often accomplished using a gradient method. The images may then be thresholded to show only marker positions against a uniform continuous background. Schinagl et al. used NIH Image 1.44 to enhance digital images obtained by CCD camera. The images were smoothed and marker edges were enhanced by convolution with a 3×3 sharpening filter and a 9×9 "Mexican Hat" filter.[93]

To increase the accuracy of displacement measurements, centroid algorithms have been developed to more accurately determine marker positions.[46,48,94,97,99] Centroid algorithms can improve accuracy from 0.5 pixels to as few as 0.02 pixels.[96] These algorithms define the spot center as the centroid of the shaded region ($\bar{x}, \bar{y}$) given by

$$\bar{x} = \frac{\sum_i \sum_j i \cdot \left(GL_{ij} - T\right)}{\sum_i \sum_j \left(GL_{ij} - T\right)} \tag{5.18}$$

and

$$\bar{y} = \frac{\sum_i \sum_j j \cdot \left(GL_{ij} - T\right)}{\sum_i \sum_j \left(GL_{ij} - T\right)} \tag{5.19}$$

where GL_{ij} is the gray level of a pixel located at (i, j) and T is a threshold level. The threshold level is chosen at a level above that of the background of the markers so that only pixels above the threshold value are used in the computation. A study by Sirkis and Lim concluded that spot sizes with a radius of about 5 pixels provided the most accurate spot position data when centroid algorithms were employed. Under optimum conditions, with centroid algorithms and lens distortion accounted for, they found that displacement measurements could be made with an accuracy of 0.015 pixels resulting in a measurement accuracy of 120 microstrains.[97]

A complete calibration and sensitivity analysis of any optical system are necessary to maximize accuracy. The tools used to calibrate the space should have an accuracy one order of magnitude greater than that which is desired from the system being calibrated. A few investigators have published thorough calibration strategies for use in determining the accuracy of particular optical systems.[55]

Derwin et al. reported a calibration of their single imager uniaxial strain system. Using calibration blocks, the system's sensitivity to errors in in-plane and out-of-plane translation and rotation were measured. In addition, effects of lighting optics, shutter settings, and imaging through a glass environmental chamber with and without a circulationg physiological saline bath were analyzed. Imaging through the glass and the circulating saline had no measurable effect on accuracy and accuracies between 500 and 1800 microstrains were reported.[22]

Smutz et al. reported the results of a calibration experiment to determine the static and dynamic accuracy of their system (Expert Vision System, Motion Analysis Corporation, Santa Rosa, CA) and the associated effect of marker size. This system has camera speeds of 200 Hz. Static error was defined as the measured motion of the markers when they were not moving. Dynamic error was the deviation of the motion calculated by the system from the motion measured by a reference LVDT. Five marker sizes from 0.8 to 3.2 mm, five camera distances, and seven loading rates were investigated. Results of the testing were compared by normalizing parameters to the camera field of view (CFC) (256 × 240 pixels). They found that static error was not a function of marker size (diameter varied from 1.6 to 50 pixels) and was equal to 0.6 pixels. Dynamic error was found to be 0.15 pixels and was independent of velocity. Consistent with the data published by Sirkis, markers with radii of 5 pixels were found to be more accurately located than smaller markers.[97] For tissue gauge lengths equal to 75% of the CFV, this system can resolve infinitesimal strains with accuracy of 830 microstrains.[99]

A complete calibration technique should quantify both systematic and random errors and their associated source and propagation effects. Each step of the image capture and analysis process should be evaluated. This includes the distortion effects due to tissue immersion, lens and lighting effects, image smoothing and sharpening processes, edge detection or centroid determinations, errors in the calibration of the displacement space, errors associated with fitting functions, and errors associated with differentiation to determine strain. The entire field of view should be calibrated to determine systematic errors due to lens distortion. Errors due to specimen rotation and movement within the plane of focus, and in and out of the plane of focus should also be quantified. The final calibration technique needs to mimic as closely as possible the actual experimental protocol, including the use of identical markers, testing environment, stimulated strains, and data reduction techniques.

Uniaxial strain of annulus fibrosus fibers was measured by Stokes and Greenapple.[100] Single fiber deformation was tracked in three dimensions using stereophotogrammetry. Two 35 mm camera images were used to determine the three-dimensional positions of the markers by using a direct linear transformation method. Seven points along the length of the fiber were tracked. A stretch ratio was calculated assuming a straight line between adjacent points referenced against a no-load condition. Errors in the technique were quantified by imposing rigid body rotations and translations on the fiber. Any measured strain was then treated as an error in the measurement technique. The digitizing procedure for a single test was conducted seven times to measure the repeatability of the digitizing process. The standard deviation in determining the point positions was found to be 0.05 mm resulting in a repeatability error in strain measurement of less than 1%.

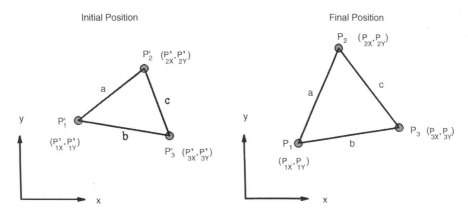

FIGURE 5.3   Initial and final positions of three points forming a triangle on the surface of a body. Lagrangian planar strains can be calculated directly from the initial and final lengths of the sides of the triangle.

Measurement of the complete strain tensor is often achieved by examining three points placed in close proximity.[7,82,111] For example, the Lagrangian strains on the surface of a tissue may be derived using the relation.

$$ds^2 - dS^2 = 2E_{ij}da^i da^j \quad i, j = 1,2 \tag{5.20}$$

where ds^2 is the deformed length, dS^2 is the undeformed length, and da^i is the change in length in the direction i. Letting P_i' and P_i denote the initial and final positions of three points undergoing a general planar deformation, as shown in Figure 5.3, the change in length of side a in the x-direction, δa_x, is given by

$$\delta a_x = (P_{3X} - P_{1X}) - (P_{3X}' - P_{1X}') \tag{5.21}$$

and in the change in length in the y-direction, δa_y, is given by

$$\delta a_y = (P_{3Y} - P_{1Y}) - (P_{3Y}' - P_{1Y}') \tag{5.22}$$

The initial length of the side a, L_a is given by

$$L_a' = (P_{3X}' - P_{1X}')^2 + (P_{3Y}' - P_{1Y}')^2)^{1/2} \tag{5.23}$$

and the deformed length of side a is given by

$$L_a = ((P_{3X} - P_{1X})^2 + (P_{3Y} - P_{1Y})^2)^{1/2} \tag{5.24}$$

Applying Eq. 5.18 directly for side a, we obtain

$$L_a^2 - L_a'^2 = 2E_{xx}(\delta a_x)^2 + 4E_{xy}\delta a_x\delta a_y + 2E_{yy}(\delta a_y)^2 \tag{5.25}$$

Using this approach for sides b and c results in three equations and the unknowns; L E_{xx}, E_{xy}, and E_{yy}, can be solved for directly. Principal strains can be calculated solving the eigenvalue problem.

With tissue property and geometry variations, uniform loads give rise to nonhomogenous strain fields. As a result, it often becomes necessary to determine the full-field strain distribution across the region of interest in the tissue. Zerniche et al. found regional surface strains near the clamp during tendon testing to be twice the value of strains in the middle of the test specimen. Further, tissue heterogeneity and the presence of an active component in muscle imply, when measuring isometric strains in a muscle tendon unit, that the strain within the structure may be changing. Van Bavel et al. simultaneously measured the strains in both the aponeurosis and muscle belly of the rat medial gastrocnemius by tracking at least three markers' displacements in the region and by directly computing the Green-Lagrange strains.[108] Trestic and Lieber reported that this relative lengthening of passive structures and shortening within the muscle belly resulted in 20% differences in predicted muscle force in the frog gastrocnemium.[106] Without regional measurements of tendon, aponeurosis, and muscle strains in their experiment, these effects might go unnoticed.

In an effort to improve accuracy and reduce noise in full-field strain measurement, investigators have fitted the surface displacement across the entire tissue surface with a function and then differentiated the function to attain the strain at each point.[8,97,101] Sutten et al. described a method which optimized smoothing parameters to remove Gaussian noise on two-dimensional displacement data.[101] Best et al. fit displacement data with a function to determine one-dimensional uniaxial finite strain in the rabbit tibialis anterior.[8] Approximately 50 marks were stained along the muscle from origin to insertion (Fig. 5.4). Image data were collected on a 1000 Hz, 238×192 pixel CCD camera. Axial deformation, u, vs. initial position of the marker on the tissue, x, was digitized for each image, resulting in a complete $u(x)$ history (Fig. 5.5). Strain was calculated using the Lagrangian formulation (Eq. 5.15). While tensile axial deformations of the muscle were large, transverse deformations and the change in transverse deformation with respect to the initial position, x, were small. At maximum displacement, dv/dx was less than 0.06; therefore, $(dv/dx)^2$ was less than 0.0036. Similarly, $(dw/dx)^2$ was less than 0.0009 at maximum displacement. Therefore, the Eq. 5.15 was simplified to

$$E_{xx}(x, t) = \frac{\partial u(t)}{\partial x} + \frac{1}{2}\left(\frac{\partial u(t)}{\partial x}\right)^2 \tag{5.26}$$

To illustrate the advantages obtained by fitting a continuous function to the displacement, derivatives of the axial displacement, $u(x)$, were determined by either a central difference method on the raw data or by differentiation of a third or fourth order polynomial fit of the displacement data (Fig. 5.6). With this particular model, the structure had a gauge length of 6 cm, and failed when elongated to 9 cm. Given the limited spatial resolution of the imager, this resulted in a position resolution of approximately 0.5 mm/pixel. Strains calculated using a central difference method from the noisy, discretized data produced errors on the order of the strain amplitude. Recognizing that discretization error is randomly distributed, fitting a function to a set of points decreases error by $1/\sqrt{n}$ where n is the number of points in the fitted curve. In these experiments, this resulted in a decrease in error by a factor of 6.3. While splines tended to have oscillations that produced negative strains when differentiated, polynomial functions were well-behaved and were insensitive to the order of the polynomial (Fig. 5.6).

When fitting functions to multiple marker positions, a tradeoff between marker size and the number of markers occurs. While both reduce error, increasing marker size decreases the number of marks which

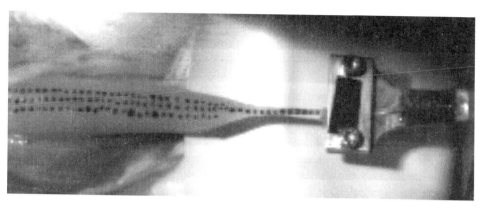

FIGURE 5.4 Digital image of the rabbit tibialis anterior with three sets of surface spots. The muscle origin is to the right. The distal tendon, left, is inserted into the grip of a hydraulic actuator. Deflections in both the lateral and axial directions were quantified by digitizing the black surface marks.

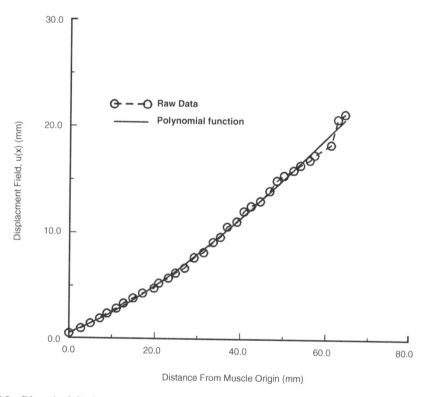

FIGURE 5.5 Discretized displacement of the surface markers on the rabbit tibialis anterior during passive elongation. These data illustrate the decrease in quantization error associated by fitting the displacement field with a polynomial function.

can be placed on the surface. As the purpose of this technique is to minimize the error in the strain field and not the position of each spot, it becomes necessary to optimize the benefits of larger spot size with the benefits of greater numbers of spots. To that end, we generated numerical strain fields from previously performed experiments to assess the effect of spot radii and number of spots on strain measurement accuracy. To determine the optimal spot size, a series of simulations was performed to emulate the CCD camera's data acquisition. Specifically, an algorithm was developed which placed a spot of a given radius

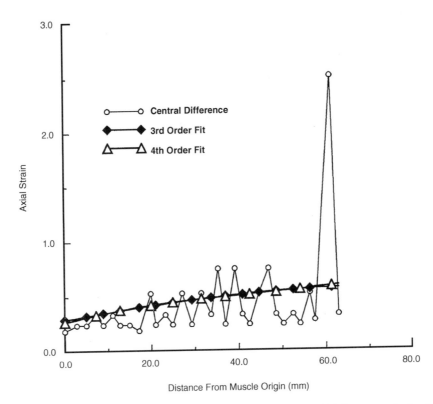

FIGURE 5.6   Axial strain calculated on the data from Figure 5.5 using a central difference method, a third order, and a fourth order polynomial to determine derivatives of the displacement field. The fitted curves are insensitive to the order of the polynomial and reduce the effects of quantization error on calculated strain by an order of magnitude.

at a random location over the pixel array. Each pixel was then assigned a gray level from 0 to 255. Based on a sample of 40 spots collected using this imaging system, the pixels completely covered by a spot were assigned a gray level of 229 ± 9.6, and the pixels that were completely uncovered were assigned a background gray level of 171 ± 6.0. Addition of the variation in pixel signal was found to profoundly influence the results, illustrating the dependence of accuracy on the unique features of the particular system under study. It also illustrates the need to calibrate and optimize each new experiment and test system.

For those pixels partially covered by a spot, a Monte Carlo routine was developed to determine the area fractions of spot and background. Gray level, GL, was then calculated based on the area fraction occupied by the spot, f, the spot's gray level, GI_s, the area fraction occupied by the background, $1 - f$, and the background gray level, GL_b, using the following equation:

$$GL = f \cdot GI_s + (1 - f) \cdot GL_b \tag{5.27}$$

Defining accuracy as the RMS error in surface strain along the length of the muscle, the effects of spot number vs. spot size were determined. Using this technique, we found that RMS strain error was minimized over a range of spot radii from approximately 2 to 7 pixels (Fig. 5.7). Because of spot distortions that occur in large spots in nonuniform strain fields, we suggest that the spot size chosen be toward the smaller end of this range.

Finite element analysis (FEA) has also been employed to determine nonhomogenous strain fields.[5,38,46,53,76,83,119] Markers placed on the surface of the specimen serve as nodes for the finite elements. The complete strain tensor can then be calculated at the Gauss points of the element. In addition, using

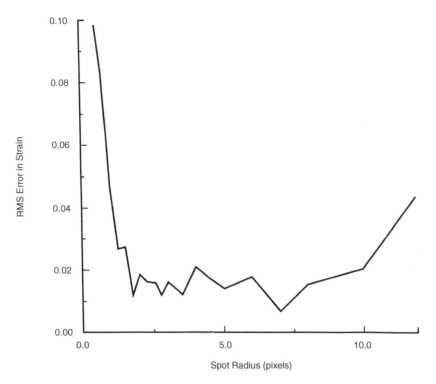

FIGURE 5.7 RMS error in strain as a function of spot radius showing that error is minimized for pixel radii of approximately 2 to 7 pixels.

interpolation functions, strain distribution throughout the element can be calculated. Specific marker placements need not be colinear nor regularly placed, and marker density can be increased in areas of high strain variation. Finite element methods may be used in both planar and three-dimensional analyses. Hoffman and Grigg used this method to determine strains within the posterior joint capsule of the cat knee using planar linear elements.[46] Hashima et al. used a least squares method and bicubic Hermite isoparametric elements to fit successive three-dimensional marker positions of the surface of a beating canine heart.[38] Sutton et al. described a penalty method used in conjunction with FEA techniques to fit noisy displacement data and determine strains.[101] Waldman and McCulloch investigated the effect of random errors introduced by Gaussian noise on the calculated strain field, finding that the FEA method reduced the errors in the strain field introduced by Gaussian noise by 50%.[113]

5.5 Conclusion

Analysis of strain in muscle has evolved considerably. Beginning with average stretches measured by simple transduction of actuator motion, to more complex and expensive optical measurement systems, and progressing to full-field, three-dimensional large strain tensorial measurement systems based on magnetic resonance and finite element techniques, the choices available to the investigator are vast. Regardless of available resources, this review demonstrates the importance of proper definition of the strain quantities to be determined and the importance of controlling the physiologic environment in which the tissue deformation is measured. It also illustrates the need to carefully design the measurement and data reduction strategy to quantify and minimize error. This is particularly true of optical systems, in which contrast, variation in contrast, spot size, and other variables all influence system accuracy, and the selection of the optimal system can only be achieved following careful experimental evaluation.

References

1. Adams, L.P., X-ray stereo photogrammetry locating the precise, three-dimensional positions of image points, *Med. Biol. Eng. Comput.*, 19, 569, 1981.
2. Anderson, J.E., Carvalho, R.S., Yen, E., and Scott, J.E., Measurement of strain in cultured bone and fetal muscle and lung cells, *In Vitro Cell Dev. Biol.*, 29A, 183, 1993.
3. Apter, J., Influence of composition on the thermal properties of tissues, in *Biomechanics: Its Foundations and Objectives*, Fung, Y.C., Perrone, N., and Anliker, M., Eds., Prentice-Hall, Englewood Cliffs, 1972, 217.
4. Arce, G.R. and McLoughlin, M.P., Theoretical analysis of the max/median filter, *IEEE Trans. ASSP-35*, 1, 60, 1987.
5. Amodio, D., Broggiato, G.B., and Salvini, P., Finite strain analysis by image processing: smoothing techniques, *Strain*, 31, 151, 1995.
6. Arms, S., Boyle, J., Johnson, R., and Pope, M., Strain measurement in the medial collateral ligament of the human knee: an autopsy study, *J. Biomechanics*, 16, 491, 1983.
7. Barbee, K.A., Macarak, E.J., and Thibault, L.E., Strain measurements in cultured vascular smooth muscle cells subjected to mechanical deformation, *Ann. Biomed, Eng.*, 22, 14, 1994.
8. Best, T.M., McElhaney, J.H., Garrett, W.E., Jr., and Myers, B.S., Axial strain measurements in skeletal muscle at various strain rates, *J. Biomechanical Eng.*, 117, 262, 1995.
9. Brown, T.D., Sigal, L., Njus, G.O., Njus, N.M., Singerman, R.J., and Brand, R.A., Dynamic performance characteristics of the liquid metal strain gage, *J. Biomechanics*, 19, 165, 1986.
10. Butler, D.L., Grood, E.S., Noyes, F.R., Zernicke, R.F., and Brackett, K., Effects of structure and strain measurement technique on the material properties of young human tendons and fascia, *J. Biomechanics*, 17, 579, 1984.
11. Butler, D.L., Noyes, F.R., and Grood, E.S., Measurements of the mechanical properties of ligaments, in *CRC Handbook of Engineering in Medicine and Biology*, Fleming, D.V. and Feinberg, B.N., Eds., CRC Press, Cleveland, 1976, Sec. B.
12. Butler, D.L., Sheh, M.Y., Stouffer, D.C., Samaranayake, V.A., and Levy, M.S., Surface strain variation in human patellar tendon and knee cruciate ligaments, *Trans. ASME*, 112, 38, 1990.
13. Butler, D.L. and Stouffer, D.C., Tension-torsion characteristics of the canine anterior cruciate ligament II. Experimental observations, *J. Biomechanical Eng.*, 105, 160, 1983.
14. Chimich, D., Shrive, N., Frank, C., Marchuk, L., and Bray, R., Water content alters viscoelastic behaviour of the normal adolescent rabbit medial collateral ligament, *J. Biomechanics*, 25, 831, 1992.
15. Cholewicki, J., Panjabi, M.M., Nibu, K., and Macias, M.E., Spinal ligament transducer based on a Hall effect sensor, *J. Biomechanics*, 30, 291, 1997.
16. Chow, G.H., LeCroy, C.M., Seaber, A.V., Ribbeck, B.M., and Garrett, W.E., Sarcomere length and maximal contractile force in rabbit skeletal muscle, *J. Orthoped. Res.*, 8, 547, 1990.
17. Comninou, M. and Yannas, I.V., Dependence of stress-strain nonlinearity of connective tissues on the geometry of collagen fibers, *J. Biomechanics*, 9, 427, 1976.
18. Crisp, J.D.C., Properties of tendon and skin, in *Biomechanics: Its Foundations and Objectives*, Fung, Y.C., Perrone, N., and Anliker, M., Eds., Prentice-Hall, Englewood Cliffs, 1972.
19. Cutts, A., The range of sarcomere lengths in the muscles of the human lower limb, *J. Anat.*, 160, 79, 1988.
20. De Clerck, N.M., Claes, V.A., Van Ocken, E.R., and Brutsaert, D.L., Sarcomere distribution patterns in single cardiac cells, *Biophys. J.*, 35, 237, 1981.
21. Demiray, H., A note on the elasticity of soft biological tissues, *J. Biomechanics*, 5, 309, 1972.
22. Derwin, K.A., Soslowsky, L.J., Green, W.D.K., and Elder, S.H., A new optical system for the determination of deformations and strains: calibration characteristics and experimental results, *J. Biomechanics*, 27, 1277, 1994.
23. Eastwood, A.B., Wood, D.S., Bock, K.L., and Sorenson, M.M., Chemically skinned mammalian skeletal muscle I. The structure of skinned rabbit psoas, *Tissue Cell*, 11, 553, 1979.

24. Elmore, S.M., Sokoloff, L., and Carmeci, P., Nature of "imperfect" elasticity of articular cartilage, *J. Appl. Physiol.*, 18, 393, 1963.

25. Figliola, R.S. and Beasley, D.E., *Theory and Design for Mechanical Measurements*, John Wiley & Sons, New York, 1991.

26. Fitzgerald, E.R., Dynamic mechanical measurements during the life to death transition in animal tissues, *Biorheology*, 12, 397, 1975.

27. Fitzgerald, E.R., Postmortem transition in the dynamic mechanical properties of bone, *Medical Phys.*, 4, 49, 1977.

28. Fleeter, T.B., Adams, P.J., Brenner, B., and Podolsky, R.J., A laser diffraction method for measuring muscle sarcomere length *in vivo* for application to tendon transfers, *J. Hand Surg. (U.S.)*, 10, 542, 1985.

29. Fung, Y.C., *Foundations of Solid Mechanics*, Prentice-Hall, Englewood Cliffs, 1965.

30. Galante, J.O., Tensile properties of the human annulus fibrosus, *Acta Orthopaed. Scand.* (suppl.), 100, 1, 1967.

31. Garrett, W.E., Jr., Nikolaou, P.K., Ribbeck, B.M., Glisson, R.R., and Seaber, A.V., The effect of muscle architecture on the biomechanical failure properties of skeletal muscle under passive extension, *Am. J. Sports Med.*, 16, 7, 1988.

32. George, D.T. and Bogen, D.K., A low-cost fiber-optic strain gauge system for biological applications, *IEEE Trans. Biomed. Eng.*, 38, 919, 1991.

33. Goldman, Y.E. and Simmons, R.M., Control of sarcomere length in skinned muscle fibres of *Rana temporaria* during mechanical transients, *J. Physiol. (London)*, 350, 497, 1984.

34. Goslow, G.E., Jr., Reinking, R.M., and Stuart, D.G., The cat step cycle: hind limb joint angles and muscle lengths during unrestrained locomotion, *J. Morphol.*, 141, 1, 1973.

35. Gottsauner-Wolf, F., Grabowski, J.J., Chao, E.Y., and An, K.N., Effects of freeze/thaw conditioning on the tensile properties and failure mode of bone-muscle-bone units: a biomechanical and histological study in dogs, *J. Orthoped. Res.*, 13, 90, 1995.

36. Graf, B.K., Vanderby, R., Jr., Ulm, M.J., Rogalski, R.P., and Thielke, R.J., Effect of preconditioning on the viscoelastic response of primate patellar tendon, *Arthroscopy*, 10, 90, 1994.

37. Hardy, W.N., Foster, C.D., Tashman, S., and King, A.I., Current findings on the kinematics of brain injury, in *Injury Prevention through Biomechanics: Symposium Proceedings*, Centers for Disease Control, Atlanta, 1997, 137.

38. Hashima, A.R., Alistair, A.Y., McCulloch, A.D., and Waldman, L.K., Nonhomogeneous analysis of epicardial strain distributions during acute myocardial ischemia in the dog, *J. Biomechanics*, 17, 795, 1993.

39. Haut, R.C., The influence of specimen length on the tensile failure properties of tendon collagen, *J. Biomechanics*, 19, 951, 1986.

40. Haut, R.C. and Little, R.W., Rheological properties of canine anterior cruciate ligaments, *J. Biomechanics*, 2, 289, 1969.

41. Haut, T.L. and Haut, R.C., The state of tissue hydration determines the strain-rate-sensitive stiffness of human patellar tendon, *J. Biomechanics*, 30, 79, 1997.

42. Haut, R.C. and Powlison, A.C., The effects of test environment and cyclic stretching on the failure properties of human patellar tendons, *J. Orthoped. Res.*, 8, 532, 1990.

43. Hawkins, D. and Bey, M., A comprehensive approach for studying muscle-tendon mechanics, *J. Biomechanical Eng.*, 116, 51, 1994.

44. Hawkins, D.A., Gomez, M.A., and Woo, S.L.-Y., An indirect method to determine ligament stresses *in situ*, *ASME Adv. Bioeng.*, 19, 166, 1986.

45. Heckman, J.L., Garvin, L., Brown, T., Stevenson-Smith, W., Santamore, W.P., and Lynch, P.R., Biplane ventriculography in the rat, *Am. J. Physiol.*, 250, H131, 1986.

46. Hoffman, A.H. and Griff, P., A method for measuring strains in soft tissue, *J. Biomechanics*, 26, 19, 1984.

47. Hughes, M.A., Myers, B.S., and Schenkman, M.L., The role of strength in rising from a chair in the functionally impaired elderly, *J. Biomechanics*, 29, 1509, 1996.
48. James, M.R., Morris, W.L., and Cox, B.N., A high accuracy automated strain-field mapper, *Experimental Mech.*, 60, March 1990.
49. Joshi, R.B., Bayoumi, A.E., and Zbib, H.M., The use of digital processing in studying stretch-forming sheet metal, *Experimental Mech.*, 117, June 1992.
50. Kahn-Jetter, Z.L. and Chu, T.C., Three-dimensional displacement measurements using digital image correlation and photogrammetric analysis, *Experimental Mech.*, 10, March 1990.
51. Kahn-Jetter, Z.L., Turso, J.A., and Pritchard, P.J., Deformed surface curve measurements using photogrammetric techniques, *Experimental Mech.*, 43, January/February 1992.
52. Kearsley, E.A., Strain invariants expressed as average stretches, *J. Rheology*, 33, 757, 1989.
53. Kraitchman, D.L., Alistair, A.Y., Chang, C.N., and Axel, L., Semi-automatic tracking of myocardial motion in MR tagged images, *IEEE Trans. Medical Imaging*, 14. 3. 1995.
54. Lai, W.M., Rubin, D., and Kermple, E., *Introduction to Continuum Mechanics*, Pergamon Press, Tarrytown, NY, 1993.
55. Lam, T.C., Frank, C.B., and Shrive, N.G., Calibration characteristics of a video dimension analyser (VDA) system, *J. Biomechanics*, 25, 1227, 1992.
56. Lanir, Y., A structural theory for the homogeneous biaxial stress-strain relationships in flat collagenous tissues, *J. Biomechanics*, 12, 423, 1979.
57. Lanir, Y., Constitutive equations for fibrous connective tissues, *J. Biomechanics*, 16, 1, 1983.
58. Leitschuh, P.H., Doherty, T.J., Taylor, D.C., Brooks, D.E., and Ryan, J.B., Effects of postmortem freezing on tensile failure properties of rabbit extensor digitorum longus muscle-tendon complex, *J. Orthoped. Res.*, 14, 830, 1996.
59. Lieber, R.L. and Friden, J., Muscle damage is not a function of muscle force but active muscle strain, *J. Appl. Physiol.*, 74, 520, 1993.
60. Lieber, R.L., Leonard, M.E., Brown, C.G., and Trestik, C.L., Frog semitendinosis tendon load-strain and stress-strain properties during passive loading, *Am. J. Physiol.*, 261, C86, 1991.
61. Lieber, R.L., Roos, K.P., Lubell, B.A., Cline, J.W., and Baskin, R.J., High-speed digital data acquisition of sarcomere length from isolated skeletal and cardiac muscle cells, *IEEE Trans. Biomed. Eng.*, 30, 50, 1983.
62. Lieber, R.L., Yeh, Y., and Baskin, R.J., Sarcomere length determination using laser diffraction: effect of beam and fiber diameter, *Biophys. J.*, 45, 1007, 1984.
63. Lipscomb, K., Cardiac dimensional analysis by use of biplane cineradiography: description and validation of method, *Cathet. Cardiovasc. Diagn.*, 6, 451, 1980.
64. Luo, P.f., Chao, Y.J., Sutton, M.A., and Peters, W.H., III, Accurate measurement of three dimensions in deformable and rigid bodies using computer vision, *Experimental Mech.*, 123, June 1993.
65. Magid, A. and Law, D.J., Myofibrils bear most of the resting tension in frog skeletal muscle, *Science*, 230, 1280, 1985.
66. McElhaney, J.H., Dynamic response of bone and muscle tissue, *J. Appl. Physiol.*, 21, 1231, 1966.
67. McElhaney, J.H., Paver, J.G., and McCrackin, H.J., Cervical spine compression responses, Paper 831615, Society of Automotive Engineers, 1983, 63.
68. Meglan, D., Berme, N., and Zuelzer, W., On the construction, circuitry and properties of liquid metal strain gages, *J. Biomechanics*, 21, 681, 1988.
69. Meglan, D., Berme, N., Zuelzer, W., and Colvin, J., Direct measurement of anterior cruciate ligament lengthening due to external loads, *ASME Adv. Bioeng.*, 19, 170, 1986.
70. Mendis, K.K., Stalnaker, R.L., and Advani, S.H., A constitutive relationship for large deformation finite element modeling of brain tissue, *J. Biomechanical Eng.*, 117, 279, 1995.
71. Merrill, T., Goldsmith, W., and Deng, Y.C., Three-dimensional response of a lumped parameter head-neck model due to impact and impulsive loading, *J. Biomechanics*, 17, 81, 1984.

72. Metz, C.E. and Fencil, L.E., Determination of three-dimensional structure in biplane radiography without prior knowledge of the relationship between the two views: theory, *Medical Phys.*, 16, 45, 1989.

73. Miles, A.W. and Tanner, K.E., *Strain Measurement in Biomechanics*, Chapman & Hall, London, 1992.

74. Morgan, D.L., Claflin, D.R., and Julian, F.J., Tension as a function of sarcomere length and velocity of shortening in single skeletal muscle fibres of the frog, *J. Physiol.* (London), 441, 719, 1991.

75. Moss, R.L., The effect of calcium on the maximum velocity of shortening in skinned skeletal muscle fibres of the rabbit, *J. Muscle Res. Cell Motility*, 3, 295, 1982.

76. Moulton, J.M., Creswell, L.L., Actis, R.L., Myers, K.W., Vannier, M.W., Szabo, B.A., and Pasque, M.K., An inverse approach to determining myocardial material properties, *J. Biomechanics*, 28, 935, 1995.

77. Myers, B.S., Woolley, C.T., Slotter, T.L., Garrett, W.E., and Best, T.M., Engineering stress-large strain constitutive behavior of skeletal muscle, *J. Biomechanical Eng.*, 120(1), 126, 1998.

78. Natori, R., The property and contraction process of isolated myofibrils, *Jikeikai Med. J.*, 1, 119, 1954.

79. Nightingale, R.W., McElhaney, J.H., Richardson, W.J., and Myers, B.S., Dynamic responses of the head and cervical spine to axial impact loading, *J. Biomechanics*, 29, 307, 1996.

80. Noonan, T.J., Best, T.M., Seaber, A.V., and Garrett, W.E., Jr., Thermal effects on skeletal muscle tensile behavior, *Am. J. Sports Med.*, 21, 517, 1993.

81. O'Dell, W.G., Moore, C.C., Hunter, W.C., Zerhouni, E.A., and McVeigh, E.R., Three-dimensional myocardial deformations: calculation with displacement field fitting to tagged MR images, *Radiology*, 195, 829, 1995.

82. Omens, J.H., MacKenna, D.A., and McCulloch, A.D., Measurement of strain and analysis of stress in resting rat left ventricular myocardium, *J. Biomechanics*, 26, 665, 1993.

83. Oomens, C.W.J., Ratingen, M.R., Janssen, J.D., Kok, J.J., and Hendricks, M.A.N., A numerical-experimental method for a mechanical characterization of biological materials, *J. Biomechanics*, 26, 617, 1993.

84. Perng-Fei, G., Strain energy function for biological tissues, *J. Biomechanics*, 3, 547, 1970.

85. Prinzen, T.T., Arts, T., Prinzen, F.W., and Reneman, R.S., Mapping of epicardial deformation using a video processing technique, *J. Biomechanics*, 19, 263, 1986.

86. Reuben, J.P., Brandt, P.W., Berman, M., and Grundfest, H., Regulation of tension in the skinned crayfish muscle fiber I. Contraction and relaxation in the absence of Ca (pCa is greater than 9), *J. Gen. Physiol.*, 57, 385, 1971.

87. Reuben, J.P., Brandt, P.W., and Grundfest, H., Tension evoked in skinned crayfish muscle fibers by anions, pH, and drugs, *J. Gen. Physiol.*, 50, 250, 1967.

88. Riemersma, D.J. and Schamhaedt, H.C., The Cryo Jaw, a clamp designed for *in vitro* rheology studies of horse digital flexor tendons, *J. Biomechanics*, 15, 619, 1982.

89. Riemersma, D.J. and van den Bogert, A.J., A method to estimate the initial length of equine tendons, *Acta Anat.* (Basel), 146, 120, 1993.

90. Rigby, B.J., Hirai, N., Spikes, J.D., and Eyring, H., The mechanical properties of rat tail tendon, *J. Gen. Physiol.*, 43, 265, 1959.

91. Sasaki, N. and Odajima, S., Elongation mechanism of collagen fibrils and force-strain relations of tendon at each level of structural hierarchy, *J. Biomechanics*, 29, 1131, 1996.

92. Sasaki, N. and Odajima, S., Stress-strain curve and Young's modulus of a collagen molecule as determined by the X-ray diffraction technique, *J. Biomechanics*, 29, 655, 1996.

93. Schinagl, R.M., Ting, M.K., Price, J.H., Gough, D.A., and Sah, R.L., Video microscopy to quantitate the inhomogeneous strain within articular cartilage during confined compression, *ASME Adv. Bioeng.*, 26, 303, 1993.

94. Sevenhuijsen, P.J., Sirkis, J.S., and Bremand, F., Current trends in obtaining deformation data from grids, *Experimental Mech.*, 22, May/June 1993.

95. Sharkey, N.A., Smith, T.S., and Lundmark, D.C., Freeze clamping musculo-tendinous junctions for *in vitro* simulation of joint mechanics, *J. Biomechanics*, 28, 631, 1995.

96. Sirkis, J.S., System response to automated grid methods, *Opt. Eng.*, 29, 1485, 1990.

97. Sirkis, J.S. and Lim, T.J., Displacement and strain measurement with automated grid methods, *Experimental Mech.*, 382, December 1991.

98. Smith, W.M. and Starmer, C.F., Error propagation in quantitative biplane cineroentgenography, *Phys. Med. Biol.*, 23, 677, 1978.

99. Smutz, W.P., Drexler, M., Berglund, E., Growney, E., and An, K.N., Accuracy of a video strain measurement system, *J. Biomechanics*, 29, 813, 1996.

100. Stokes, I. and Greenapple, D.M., Measurement of surface deformation of soft tissue, *J. Biomechanics*, 18, 1, 1985.

101. Sutton, M.A., Turner, J.L., Bruck, H.A., and Chae, T.A., Full-field representation of discretely sampled surface deformation for displacement and strain analysis, *Experimental Mech.*, 168, June 1991.

102. Taylor, R.G., Geesink, G.H., Thompson, V.F., Koohmararie, M., and Goll, D.E., Is Z-disk degeneration responsible for postmortem tenderization? *J. Anim. Sci.*, 73, 1351, 1995.

103. Tencer, A.F. and Ahmed, A.M., The role of secondary variables in the measurement of the mechanical properties of the lumbar intervertebral joint. *J. Biomechanical Eng.*, 103, 129, 1981.

104. Tidball, J.G. and Daniel, T.L., Elastic energy storage in rigored skeletal muscle cells under physiological loading conditions, *Am. J. Physiol.*, 250, R56, 1986.

105. Tidball, J.G., Salem, G., and Zernicke, R., Site and mechanical conditions for failure of skeletal muscle in experimental strain injuries, *J. Appl. Physiol.*, 74, 1280, 1993.

106. Trestik, C.L. and Lieber, R.L., Relationship between Achilles tendon mechanical properties and gastrocnemius muscle function, *J. Biomechanical Eng.*, 115, 225, 1993.

107. Valanis, K.C. and Landel, R.F., The strain-energy function of a hyperelastic material in terms of the extension ratios, *J. Appl. Phys.*, 38, 2997, 1967.

108. van Bavel, H., Drost, M.R., Wielders, J.D., Huyghe, J.M., Huson, A., and Janssen, J.D., Strain distribution on rat medial gastrocnemium (MG) during passive stretch, *J. Biomechanics*, 29, 1069, 1996.

109. van Weeren, P.R., Jansen, M.O., van den Bogert, A.J., and Barneveld, A., A kinematic and strain gauge study of the reciprocal apparatus in the equine hind limb., *J. Biomechanics*, 25, 1291, 1992.

110. Veronda, d.R. and Westman, R.A., Mechanical characterization of skin-finite deformations, *J. Biomechanics*, 3, 111, 1970.

111. Villarreal, F.J., Waldman, L.K., and Lew, W.Y., Technique for measuring regional two-dimensional finite strains in canine left ventricle, *Circ. Res.*, 62, 711, 1988.

112. Waldman, L.K., Fung, Y.C., and Covell, J.W., Transmural myocardial deformation in the canine left ventricle: normal *in vivo* three-dimensional finite strains, *Circ. Res.*, 57, 152, 1985.

113. Waldman, L.K. and McCulloch, A.D., Nonhomogeneous ventricular wall strain: analysis of errors and accuracy, *J. Biomechanical Eng.*, 115, 497, 1993.

114. Winter, D.A., *Biomechanics and Motor Control of Human Movement*, John Wiley & Sons, New York, 1990.

115. Wollschlager, J., Lee, P., Zeiher, A., Solzbach, U., Bonzel, T., and Just, H., Derivation of spatial information from biplane multidirectional coronary angiograms, *Med. Prog. Technol.*, 11, 57, 1986.

116. Woo, S.L.Y., An, K.N., Arnoczky, S.P., Wayne, J.S., Fithian, D.C., and Myers, B.S., Anatomy, biology, and biomechanics of tendon, ligament, and meniscus, in *Orthopaedic Basic Science*, Simon, S.R., Ed., American Academy or Orthopaedic Surgeons, 1994.

117. Woo, S.L., Ritter, M.A., Amiel, D., Sanders, T.M., Gomez, M.A., Kuei, S.C., Garfin, S.R., and Akeson, W.H., The biomechanical and biochemical properties of swine tendons: long-term effects of exercise on the digital extensors, *Connective Tissue Res.*, 7, 177, 1980.

118. Yin, F.C., Tompkins, W.R., Peterson, K.L., and Intaglietta, M., A video-dimension analyzer, *IEEE Trans. Biomed. Eng.*, 19, 376, 1972.

119. Young, A.A., Kraitchman, D.L., Dougherty, L., and Axel, L., Tracking and finite element analysis of stripe deformation in magnetic resonance tagging, *IEEE Trans. Medical Imaging*, 14(3), 413, 1995.
120. Zajac, F.E., Muscle and tendon: properties, models, scaling, and application to biomechanics and motor control, *Crit. Rev. Biomed. Eng.*, 17, 359, 1989.
121. Zernicke, R.F., Butler, D.L., Grood, E.S., and Hefzy, M.S., Strain topography of human tendon and fascia, *J. Biomechanical Eng.*, 106, 177, 1984.

6

A Review of the Technologies and Methodologies Used to Quantify Muscle-Tendon Structure and Function

David Hawkins
University of California at Davis

6.1   Introduction

Muscle-tendon units are complex biological actuators able to generate considerable force to stabilize and/or move segments of the body and absorb energy imparted to the body. They are controlled through neural inputs and generate their forces by converting chemical energy into mechanical energy. Their mechanical behavior is directly linked to their macroscopic and microscopic structures and the properties of the specific proteins constituting these structures. Muscle-tendon units are highly adaptable, modifying their structure and protein forms in response to changes in environmental stimuli. Due to the integral role skeletal muscle plays in human function, an understanding of its behavior has been of interest for thousands of years. However, because of its complex organization of membranes, organelles, proteins, nerves, and vessels, and its versatility and adaptability, increases in our understanding of the detailed workings of skeletal muscle have often depended on the development of new technologies and methodologies. Much is still unknown about muscle-tendon structure and function and it is likely that further knowledge in this area will be achieved through technological innovations.

The purpose of this chapter is to provide detailed descriptions of muscle-tendon structure and function, and to summarize many of the technologies and methodologies employed over the years to unravel the intricate structures and functions of muscle-tendon units. While structure and function are directly related, for the sake of simplicity, they will be discussed separately. Muscle-tendon structure will be presented first, and a review of various approaches employed to study this structure will follow. Muscle-tendon function will be presented next, followed by a review of the approaches employed to study function.

6.2   Muscle-Tendon Structure

In this section, a detailed description of the structural organization of a muscle-tendon unit is presented. The description of the structural organization of muscle begins at the level of the whole muscle and proceeds to the smaller subunits, concluding with the proteins constituting the myofilaments. Membrane systems, neural, vascular, and connective tissue networks are described. The variability in muscle fiber structures and how this variability has led to various fiber-type naming schemes will then be discussed.

Skeletal muscle exists in a variety of shapes and sizes. It is composed of many subunits arranged in an organized, but complex manner (see Fig. 6.1). Additionally, muscles connect in series to tendons, are innervated by nerves, and supplied with vascular networks. A whole muscle is surrounded by a strong sheath called the epimysium, and divided into a variable number of subunits called fasciculi. Each fasciculus is surrounded by a connective tissue sheath called the perimysium. Fascicles may be further divided into bundles of fibers (or muscle cells) surrounded by a connective tissue sheath called the endomysium.[8,26,51,54,88,91,108,109,110] Beneath the endomysium are two additional membranes, the basal lamina and the plasmalemma.[26,88,96] The orientation of fibers relative to the line of action of the muscle-tendon complex is referred to as the pinnation angle. In humans, the pinnation angle ranges from 0 to 25°.[88,121] Muscle may be classified as fusiform (or spindle), penniform, bipenniform, triangular, rectangular (or strap), and rhomboidal. Fibers attach at both ends to tendon or other connective tissue. Muscle fibers contain mitochondria, multiple nuclei, ribosomes, soluble proteins, lipids, glycogen, and satellite cells. Fibers are cylindrical, with their diameter ranging from 10 micrometers (μm) to 100 μm (smaller than the size of a human hair).[88] They may be a few millimeters (mm) or many centimeters (cm) in length. Fibers are subdivided radially into myofibrils having diameters of approximately 1 μm. Myofibrils are divided longitudinally into sarcomeres and radially into myofilaments. A saromere is defined as the region between Z-lines (defined below). Sarcomeres have a rest length of about 2.0 to 3.0 μm. Myofilaments are often classified as either thick or thin filaments.

Thick filaments are composed primarily of myosin molecules. Myosin accounts for approximately 55% of the myofibril volume. It is composed of two heavy chains and four light chains. Two light chains are associated with each heavy chain. The two heavy chains are identical, whereas the light chains vary within different fiber types. Each myosin molecule is rod shaped with two adjacent globular heads at one end. The myosin molecule structure has been defined in terms of two general regions: the light meromyosin (LMM), and the heavy meromyosin (HMM). The LMM represents part of the tail. The HMM contains the two heads, and the remaining part of the tail not considered part of the LMM. HMM is further divided into subfragment 1 (S1) and subfragment 2 (S2) (see Fig. 6.1). Myosin molecules are about 160 nanometers (nm) long (myosin rod is 140 nm and head is 15 nm) and 2 nm in diameter.[8,26,108,110] Myosin molecules are arranged to give a total thick filament length of 1.55 μm and 12 to 15 nm diameter.[80] There are approximately 100 axial locations along the thick filament, separated by 14.3 nm where myosin heads exist. The number of myosin molecules terminating at each axial repeat location is still controversial. Most of the evidence has been interpreted as suggesting three myosin ends per axial repeat distance. Each

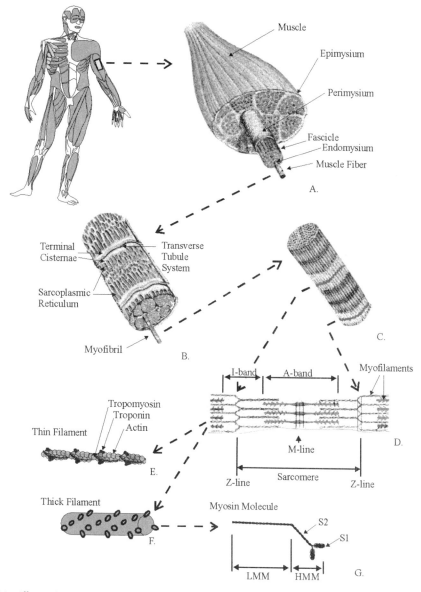

FIGURE 6.1 Illustration of the structutral organization of muscle. A whole muscle is shown in A, a muscle fiber in B, a myofibril in C, a sarcomere in D, a thin filament in E, a thick filament in F, and a myosin molecule in G.

thick filament contains approximately 300 myosin molecules (assuming three myosin ends per axial repeat location).[26] At least 8 proteins in addition to myosin are affiliated with the thick filament: C-protein, H-protein, M-protein, myomesin, M-creatine kinase, adenosine monophosphate (AMP) deaminase, skelemin, and titin.[8,26,88,110]

Thin filaments are composed primarily of actin, tropomyosin, and troponin. Thin filaments are approximately 1 μm long and 8 nm in diameter. Each thin filament contains about 360 actin monomers. Each actin monomer consists of a single polypeptide chain.[8] Actin monomers polymerize to form a double helix pattern with a repeat spacing of 5.5 nm.[8,88] Because of symmetry and the spherical shape of the actin monomers, there exists a groove on either side of the helix chain. Each groove is filled by a series of tropomyosin-troponin complexes, each spanning a length of seven actin monomers (41 nm in length). There is one troponin molecule, approximately 26 nm long, for each tropomyosin molecule.

The tropomyosin molecule forms an α-helical coiled coil structure. The troponin molecule can be further divided into troponins C, I, and T.[88,108]

Thick and thin filaments are oriented parallel to one another within a sarcomere and typically have a zone of overlap (see Fig. 6.1). The region containing the thick filaments is referred to as the anisotropic or A-band, approximately 1.55 μm in length. The region containing the thin filaments with no overlap with the thick filaments is termed the isotropic or I-band. The 0.16 μm region in the center of the A-band that has no thin filament overlap is called the Helle* or H-zone. In the middle of the A-band is a region called the middle or M-line. The M-line is composed of a connective tissue network binding the thick filaments. At the end of each sarcomere is a dense protein zone called the Z-line** (also referred to as the Z-disk or Z-band).[42,91] The Z-disk is composed of a connective tissue network binding the thin filaments. It contains the proteins α-actinin, desmin, filamin, and zeugmatin.[26] Thin filaments are attached at the Z-disk but are free to interdigitate with the thick filaments at their other ends. When viewed in cross section through the zone of overlap between thin and thick filaments, a hexagonal lattice appears with one thick filament surrounded by six thin filaments. The spacing between thick filaments is 40 to 50 nm.[80] The spacing between thick and thin filaments is 20 to 30 nm.[8]

The muscle fiber contains two distinct membranous systems: the transverse tubular system (T-system or T-Tubule system) and the sarcoplasmic reticulum (SR) (see Fig. 6.1).[8,26,80,88] The T-system is part of the plasmalemma and makes a network of invaginations into the cell near the Z-line in amphibian muscle and near the junction of the A- and I-bands in mammalian muscle.[26] No part of the contractile machinery is further than 1.5 μm from a T-tubule.[72] Two terminal cisternae (part of the SR) run parallel to the T-system to form a triad.[96] The T-system is separated from the terminal cisternae by a distance of about 16 nm but connects to the terminal cisternae via numerous feet.[72] The SR traverses longitudinally from the terminal cisternae.

In addition to the structures mentioned above, vascular, neural, and connective tissues play important roles in muscle function. Muscles have a rich supply of blood vessels that supplies the oxygen needed for oxidative metabolism. Capillary networks are arranged around each fiber with the capillary densities varying around different fiber types.[80]

The basic neuromuscular element is called the motor unit. It consists of a single alpha motoneuron and all the muscle fibers it innervates. The number of fibers per motor unit is variable, ranging from just a few in ocular muscles requiring fine control, to thousands in large limb muscles.[23,80] Fibers from a given motor unit tend to be dispersed throughout the muscle cross section rather than clumped together in one region. Oxidative fibers tend to occur in greater percentages deeper in the muscle compared to glycolytic fibers which have higher percentages in the perphery.[89] The structure of the neuromuscular junction can vary significantly between different species, between different fiber types of the same species, and during the course of development. In general, the nerve terminal ending on a muscle fiber contains vesicles 50 to 60 nm in diameter. These vesicles contain acetylcholine (Ach), adenosine triphosphate (ATP), a vesicle-specific proteoglycan, and a membrane phosphoprotein, synapsin. Approximately 15% of the nerve terminal volume is taken up by mitochondria. The nerve and muscle membranes are not in direct contact. The synaptic space is approximately 50 to 70 nm wide and contains acetylcholinesterase (AchE). The muscle membrane contains nicotinic Ach receptors.[26] The muscle membrane has several folds in the regions of the nerve endings to increase the transmitter reception area eightfold to tenfold.

Muscles have extensive connective tissue networks located both in parallel and in series with the fibers. Myofibrils appear to be attached transversely at periodic adhesion sites. The protein titin spans the distance between Z-lines and the middles of the thick filaments.[8] Muscle fibers are connected in series with tendons. The primary structural unit of tendon is the collagen molecule. Type I collagen consists of three polypeptide chains coiled together in a right-handed triple helix held together by hydrogen and covalent bonds.[43,120] Collagen molecules are organized into long, cross-striated fibrils that are arranged into bundles to form fibers. Fibers are further grouped into bundles called fascicles, which group together

---

*German for "light."

**From Zwischen-Scheibe, meaning "interimdisk."

to form the gross tendon. Elastic and reticular fibers are also found in tendon along with ground substance (a composition of glycosaminoglycans and tissue fluid). In an unstressed state, collagen fibers take on a sinusoidal appearance, referred to as a crimp pattern.

Although the general structures (i.e., actin and myosin filament lengths and their lattice arrangement) are similar among vertebrate muscle fibers, there are differences in the regulatory proteins of the myosin and troponin, the extensiveness of membrane networks, and the number of mitochondria and other organelles. These variations have functional consequences that led to the development of a variety of naming schemes to identify fibers with specific structural and functional properties (e.g., red/white, fast/slow, oxidative/glycolytic, types I/IIa,b,c, and SO/FOG/FG).[19,20,23-25,29,94,107] The myosin molecule appears in various isoforms.[56,79,105] These isoforms exhibit different amino acid sequences, ATPase activity, and affinity for calcium.[99] The troponin C protein may vary in its sensitivity to calcium. There are differences in the membrane networks. The T-system may be twice as extensive in one fiber compared to another. Mitochondrial density also varies among fibers.[26]

6.3   Approaches Used to Study Muscle-Tendon Structure

Our understanding of the complex structural organization of muscle-tendon units described above has come from keen observations and the development of a variety of technical tools and novel methodologies. The first recorded scientific medical studies were undertaken by the Greeks around the 6th century B.C.[9] However, most of the studies conducted prior to the 17th century, which contributed to our understanding of muscle structure, were based on gross dissections and involved identifying muscles, tendons, nerves, and the vascular network. Since then, advances in mathematics, chemistry, physics, and genetics have played a major role in identifying and characterizing muscle-tendon structure.

Microscopy has been used extensively to study muscle. Lenses were first used to magnify objects around 1600 A.D.[104] Microscopes, in which various arrangements of flat, concave, and convex lenses are used to magnify images, were introduced around the beginning of the 17th century. Microscopy has developed into a highly technical field utilizing a variety of illuminating approaches.

Light microscopy was the first technique employed to study muscles and other biological tissues. Leeuwenhoek (1632–1723) was one of the first great biological microscopists. He manufactured hundreds of microscopes which he used to observe many biological tissues. Unfortunately, much of his expertise in tissue preparation and illumination was lost throughout the 18th and 19th centuries. Much of the work in light microscopy conducted then centered around correcting for artifacts and aberrations through matching glass, refractive media, and improving lens manufacturing.[104] Muscle appears transparent when viewed using normal light microscopy, and therefore it is often stained prior to viewing. A variety of stains have been used to provide the contrast necessary to identify different organelles and gross structures.[104] In addition, the light used to illuminate the specimen has been manipulated in various ways to cause refraction and interference patterns that allow different structures within muscle to be visible.

Dark-ground, phase contrast, interference, and polarization microscopy identify regions of different refractive indices, but they accomplish this based on fundamentally different approaches. While most living, non-stained biological tissue is transparent when investigated with normal light microscopy, different regions of a cell have different refractive indices. In dark-ground microscopy, light is passed through the specimen at rather oblique angles so that the direct light beam passes to the side of the objective.[104,114] The only light entering the objective comes from refracted light. Regions of high refractive index appear bright against a black background as they reflect the light to the eyepiece or viewing port. Phase contrast microscopy makes use of the relative phase differences in light passing through different regions of the tissue having different refractive indices. These phase differences are converted to changes in light intensity in the image plane.[114] Interference microscopy splits the illuminating beam into two beams. One beam passes through the specimen and the other beam passes around it.[8] The two beams are recombined before the objective. Light passing through high refractive index tissue is slowed down, phase shifted, relative to light passing around the tissue. The interference pattern that results indicates different protein-dense zones. If the proteins within a region which give rise to its refraction index are not homogeneously distributed, then the refractive index will depend on the plane of polarization of light. A polarization microscope takes advantage of this property. Basically, a polarizer located at the condenser causes a single plane of light to illuminate the specimen. An analyzer located after the specimen allows a single plane of light to pass to the objective. The alignment of polarizer and analyzer is variable, but they are usually set at right angles.[104,114] The object stage can rotate relative to the plane of polarization. The terminology commonly used to describe sarcomere anatomy is largely the result of muscle observations made under polarization microscopes. When viewed with a polarization microscope, specific zones of a muscle fiber appear darker than other zones. The dark zones have dense protein bands causing the plane of polarization of light to be strongly rotated. These zones have been labeled anisotropic or A-bands. Other zones are less protein dense and rotate the plane of polarization of light weakly. These zones have been labeled isotropic or I-bands.[8,51] The Z-band is also observed to be anisotropic while the H-zone in the middle of the A-band appears relatively isotropic.

The use of light as an illuminating medium has inherent resolution limitations. Basically, the best resolving power of a microscope is equal to about 0.6 times the wavelength of the electromagnetic radiation used to illuminate the specimen. The use of short wavelengths provides better resolution (e.g., 475 nm wavelength blue light provides better resolution than 700 nm wavelength red light, and X-rays with wavelengths of about 0.1 nm are better than visible light). The attainable resolving power of light microscopy is about 200 nm and that of electron microscopy is about 0.1 nm.[104] Based on the various structural dimensions presented previously, it is evident that light microscopy could be used to distinguish Z-lines with 2 to 3 μm separation distances, but could not be used to distinguish between myofilaments having spacings of 20 to 50 nm.

Due to resolution limitations inherent in using light, further resolution of muscle structure using microscopy depended on the development of electron microscopy (EM). The theoretical concept of an electron microscope was proposed in the 1920s.[104] The concept was formulated from the ideas that particles have wave properties and a magnet can be used to focus a beam of electrons similar to the way a lens focuses light. By the 1940s many countries were making transmission electron microscopes. Following the development of transmission electron microscopy (TEM), scanning electron microscopy (SEM) was developed. SEM utilizes the reflected electrons to make an image of the object in contrast to recording the transmitted electrons in TEM. It has the advantage of providing greater topographical information about the specimen than TEM. However, SEM provides a very low contrast signal, and its utility has relied on the development of computer algorithms for amplifying, averaging, and processing the signals in other ways.

Conventional preparation of a specimen for EM involves fixation by cross-linking agents, dehydration, embedding in resin, sectioning, and staining with electron-dense heavy metals. One obvious drawback to this technique is that the tissue is dead and harshly handled prior to viewing. Nonetheless, electron microscopy has revealed much about muscle and tendon structure. It revealed that the banding pattern in skeletal muscle arises from interdigitation of sets of filaments. Thin filaments were observed to connect to the Z-line and make up the I-band. Thick filaments were observed to compose the A-band with thick and thin filaments having a region of overlap. High magnification electron micrographs showed connections between thick and thin filaments in the overlap zone. These connections were referred to as cross-bridges. EM, in combination with techniques such as freeze-fracture and protein purification, has provided much of what we know about the structure of contractile proteins, the membrane networks, and the neural innervation zones.[8,26,108]

In addition to microscopy, muscle has been examined using diffraction techniques. A diffraction pattern arises whenever a beam of electromagnetic radiation passes through a narrow slit or a small hole. The hole or slit causes the beam to spread and acquire regions of destructive interference such that a banding pattern or a series of concentric rings results. When monochromatic light is used to illuminate muscle, the striation pattern within muscle gives rise to an optical diffraction pattern. The distance between fringes can be used to calculate sarcomere length.[8] X-rays having wavelengths of about 0.1 nm can be used to illuminate muscle and create a diffraction pattern that can be used to calculate the spacing between filaments, the spacing between cross-bridges, and even the spacing between actin monomers

(5.5 nm).[8,88,110] This technique in conjunction with EM has been used extensively to reveal much of what we know about the molecular structure of muscle. A major advantage of diffraction studies is that they can be applied to thin sections of living tissues.

A variety of other techniques have been used to identify the molecular structure of muscle. Thick and thin filament composition were determined through extraction/aggregation studies. Selective extraction of A- and I-bands with salt solutions revealed that thick filaments are composed mainly of myosin and thin filaments are composed mainly of actin. Evidence indicating that the cross-bridges represent the HMM end of myosin came from aggregation studies.[109] When LMM aggregated it gave a smooth structure. When intact myosin molecules aggregated they formed a large number of projections. Different myofibrillar isoforms have been identified using peptide finger printing, monoclonal antibodies, and the application of recombinant DNA procedures.[26] Fluorescence techniques are now used to study protein distribution within a cell.[68]

Like muscle, tendon structure has been determined using a variety of techniques. Chemical techniques have been used to determine its protein and molecular components. Light microscopy and tissue staining techniques have revealed the vascular, neural, and fiber structures within tendon as well as the locations of fibroblast cells. Polarization microscopy in combination with special stains has been used to isolate the fibrous elements of collagen, elastin, and reticulin. Electron microscopy has been used to determine the organization of collagen molecules.[43,120] A summary of some of the approaches used to study muscle-tendon structures is given in Fig. 6.2.

Summary of Approaches Used to Determine Muscle-Tendon Structures	
Approach Employed	**Examples of Structures Identified**
I. Gross Dissection	I. Muscle-tendon attachments and gross, architecture, blood vessels, nerves

II. Microscopy
A. Light
  1. Normal with stains
  2. Dark-ground
  3. Phase-contrast
  4. Interference
  5. Polarization
	II. Cell structures
	A. Microscopic cell structures
	  1. Muscle cell organelles, membranes
	  2. Regions of different refractive index
	  3. Regions of different refractive index
	  4. Regions of different refractive index
	  5. A- and I-bands, Z-lines

B. Electron
  1. TEM
  2. SEM
	B. Molecular structures
	  1. Actin and myosin, cross-bridges
	  2. 3 dimensional images of membrane vesicles and contractile proteins

III. Diffraction A. Monochromatic Light B. x-ray	III. Spacing between structures A. Sarcomere lengths B. Axial repeat spacing of myosin heads, myofilament spacing
IV. Chemical A. Extraction combined with electron microscopy B. Antibody labeling combined with electron microscopy C. Electrophoresis	IV. Chemical composition A. Contractile proteins and sub-fragments B. Contractile proteins and sub-fragments C. Molecular weight of proteins

FIGURE 6.2 A summary of various approaches that have been used to study muscle-tendon structure.

6.4   Muscle-Tendon Function

This section provides descriptions of the functions performed by the individual structures identified in the previous section, the processes involved in energy supply, the processes involved in converting chemical energy into mechanical force, and the factors that affect muscle-tendon performance.

Functions of Specific Structures

Nuclei dictate cell material and distribution. Like cell managers, they keep structures organized. Nuclei communicate with other nuclei within a cell to maintain some consistency of regulation.[88] They also exhibit local regulatory control, especially at locations near the sites of neural innervation. The amount and type of protein to be produced are defined by a nucleus and carried out by the ribosomes in response to mRNA. Ribosomes are granules of ribonucleoprotein. Protein synthesis can be up- or down-regulated fairly quickly, providing muscle the ability to adapt. The speed, strength, and endurance properties of the cell are dictated by the proteins comprising the cell.

Mitochondria located in the cytoplasm produce ATP through oxidative metabolism. ATP is the energy source used for all cell functions (e.g., protein synthesis, ion transport, repair, and force production). Mitochondrial density depends on function. It may be as high as 20% by volume for highly oxidative fibers.[41,42]

Other important substances contained in the cytoplasm are glycogen, lipids, and enzymes. Glycogen and lipids are sources of ATP. Glycogen is a polymer of linked glucose which can be used as an immediate source of ATP through anaerobic glycolysis performed by soluble enzymes. Lipids serve as a second energy source, but require oxygen for their metabolism. Thus, they are most prevalent in cells with high mitochondrial density.[88]

The extensive membrane network of muscle cells performs several functions. The endomysium provides structural support for the muscle fiber and the neural and vascular tissues interacting with it. The basal lamina appears to play a role in injury repair. Complete repair can occur rapidly if the basal lamina is intact to provide a scaffold for regeneration.[26,54,88] The basal lamina also communicates with the nerve to signal it where to innervate the muscle fiber if denervation has occurred. The plasmalemma, T-system, and SR function as semi-permeable barriers, conduits for electrical signal propagation, filters, and calcium storage centers. The plasmalemma acts as a filter by requiring a certain number of receptors on its surface to be stimulated before changing its membrane permeability and conducting the electrical signal of the nerve into the cell. The T-system provides the conduit for rapid transmission of electrical activity to the inner regions of the cell. The SR stores and releases calcium ions which are essential for force production and relaxation.

Sarcomeres are the basic units of shortening and force generation and thus have numerous structures of functional importance. The Z-line is a highly organized structure that interconnects the thin filaments in a very precise array. The M-line is presumed to be responsible for binding the thick filaments and maintaining them in a hexagonal pattern when viewed in a transverse plane. The thick filaments contain myosin molecules which perform several tasks. The HMM portion of myosin is often referred to as the cross-bridge because it is the structure that reaches out and binds to actin during contraction. The HMM-LMM interface is flexible, allowing the S1 portion of HMM to project out about 55 nm[8] to reach a thin filament. S1 contains binding sites for two light chains: ATP and actin. Thin filaments play an equally important role in force production. Actin monomers have binding sites compatible with regions of the S1 portion of myosin. These binding sites are normally covered by tropomyosin during rest conditions. However, in the presence of calcium, troponin C, which is sensitive to calcium ion binding, causes troponin I to produce a conformational change in tropomyosin which then exposes the myosin binding sites. Troponin T functions to regulate troponin-tropomyosin binding. Two final structures that may have functional importance are nebulin and titin. Nebulin runs parallel to the actin filaments and may function in length determination during assembly. Titin is a relatively large elastic filament that stretches from M-line to Z-line. It provides passive elasticity and helps to keep the A-band centralized.[8]

Processes Involved in Energy Supply

All the processes involved in cell maintenance and force production rely on the availability of ATP and thus a discussion of the processes involved in ATP synthesis and supply is relevant. ATP is the universal energy source for all cells. Energy comes from splitting ATP into adenosine diphosphate (ADP) and inorganic phosphate (Pi). ATP is normally bound to Mg in skeletal muscle, but myosin can hydrolyze ATP and release its energy. This reaction is very slow in isolation, about 0.01 ATP/sec, but in the presence of actin this rate increases to 4.5 ATP/s and in actual skeletal muscle this process proceeds at a rate of about 6.3 ATP/myosin head/sec.

The body provides several means of supplying ATP to muscle.[73,74] The amount of ATP present in living muscle can provide enough energy for only about eight muscle twitches.[91] Obviously the body provides some means of quickly replenishing ATP. The pathway most commonly used during the onset of physical activity combines ADP with phosphocreatine (PCr) to produce ATP and creatine (Cr). This reaction is often referred to as the Lohmann reaction and can take place in either direction. However, the equilibrium constant for the reaction favors the production of ATP by a factor of about 20. PCr must be present in the muscle for the Lohmann reaction to proceed toward ATP production. Muscle maintains a small reserve of PCr, but not enough to supply the amount of ATP needed for sustained activities. In fact, the amount of PCr stored in muscle tissue can provide enough ATP to sustain several hundred twitches.[8] This is much greater than what the stores of ATP can supply, but still not sufficient to supply the energy demands placed on the body during daily activities.

Aerobic phosphorylation and anaerobic glycolysis provide additional pathways for ATP production. Anaerobic glycolysis can be considered a process in itself or a precursor to oxidative phosphorylation. Whether or not oxidative phosphorylation occurs depends on oxygen availability to the muscle cell and the content of cytochromes and myoglobin present within the cell. During anaerobic glycolysis, which takes place in the cytoplasm, a series of reactions break down glucose to form two pyruvic acid, two hydrogen, and four ATP molecules. Anaerobic glycolysis utilizes two ATP molecules to breakdown glucose, hence the net yield is two ATP molecules. The pyruvic acid and hydrogen molecules generated from anaerobic glycolysis enter the mitochondria where the Kreb's cycle (also referred to as the tricarboxylic acid or TCA cycle) takes place. For each pyruvic acid molecule entering the Kreb's cycle, three CO_2 molecules, five hydrogen molecules, and one ATP molecule are formed. The hydrogen atoms released from both the Kreb's cycle and anaerobic glycolysis enter an electron transport system (ETS) by combining with nicotinamide-adenine dinucleotide (NAD). Aerobic oxidative phosphorylation will occur at this stage if sufficient oxygen is available to meet the supply of hydrogen transported to the mitochondria via NAD. If the oxygen supply is not sufficient, then NADH reacts with the pyruvic acid to form lactic acid. Lactic acid can accumulate in the muscle and cause fatigue. At some point, usually during a recovery period, the lactic acid is cleared from the muscle and carried to the liver where it is synthesized into glucose. Provided oxygen is available, a total of 32 ATP molecules along with CO_2 and water are produced from the NADH. Energy is needed to transport the two hydrogen molecules generated during anaerobic glycolysis from the cytoplasm into the mitochondria. This process utilizes one ATP molecule per hydrogen molecule transferred. Thus the net yield of ATP per glucose molecule from aerobic metabolism is 34. The aerobic processes are much more efficient than anaerobic glycolysis acting alone, which yields only two ATP molecules per glucose molecule. Also no lactic acid is formed; only CO_2 and H_2O are produced.

Processes Involved in Force Development and Transmission

Muscles generate force by converting chemical energy into mechanical force in response to electrical signals received from a motoneuron. The basic functions of force development and shortening are initiated through the processes of excitation-contraction coupling. These processes are initiated when a peripheral nerve action potential arrives at a muscle fiber's synaptic cleft (or motor end plate). This action potential may result from signals sent from the brain or through reflex pathways (discussed more in the section titled "Effects of an Integrated Multiple Muscle System"). Signals are passed from nerve to muscle by chemical transmitters. When an electrical signal arrives at a motor end plate, the membrane allows calcium to flow into the cell.[27] The increased intracellular calcium ion concentration causes vesicles located on the membrane to release acetylcholinesterase (Ach) which diffuses across the synaptic cleft and binds to specific receptors on the muscle membrane. If sufficient binding takes place, then the permeability of the muscle membrane changes (reaches threshold).[54]

The number of receptors that must be stimulated to cause these changes varies for different fiber types. Permeability changes cause sodium ions to enter the cell and potassium ions to leave the cell. The membrane depolarizes, becoming less negative inside the cell. The signal, or action potential, is propagated in both directions along the length of the muscle fiber. An action potential is always the same for a given cell. The cell depolarizes in an all-or-none response once a sufficient stimulus is achieved. After the action potential, there is a refractory period in which the cell cannot be activated again. The refractory period is necessary to prevent back flow of impulses.

Excitation of the muscle membrane spreads inward through the T-system which communicates this excitation to the SR. The SR then releases calcium ions along the length of the fiber. The calcium binds with troponin C which causes troponin I to create a conformational change in tropomysin which exposes an actin binding site for myosin.[80,96] Two calcium receptors must be stimulated in slow oxidative fibers to remove the inhibitory effect of Troponin I, while only one is required in fast glycolytic fibers. The S1 portion of a neighboring myosin molecule binds with the actin and develops force. If the force developed by all bound myosin heads is greater than the external force applied to the muscle or muscle-tendon unit, then the muscle will shorten. The muscle will lengthen or remain at a constant length if the force is less than the external force, or equal to the external force, respectively. Force will continue as long as there are bound myosin heads. However, in the presence of ATP, the myosin adenosine triphosphatase (ATPase) will hydrolyze the ATP and the acto-myosin bond will be broken. Myosin ATPase activity is approximately three times faster in fast-glycolytic fibers than it is in slow oxidative fibers.[59,86] Myosin will continue to form new bonds with actin as long as there is sufficient calcium to bind with troponin C. Once the action potential stops the Ca^{+2} is pumped back into the SR. The rates of myosin ATPase activity and membrane system release and uptake of Ca^{+2} regulate the rate of force development and relaxation.

Factors Affecting Muscle-Tendon Performance

The force developed by the muscle and actually transmitted to the bones via its associated tendons depends on the neural input, the muscle-tendon architecture, the muscle kinematics, the muscle composition of different fibers, the contraction history, and the feedback from various proprioceptors.

Effects of Neural Input

The level of force generated by voluntary contraction of skeletal muscle is controlled by at least two neural mechanisms, motor unit recruitment and modulation of the firing rate of active motor units (rate coding). It is generally accepted that motor units are recruited in an orderly manner consistent with the size principle of Henneman et al.[64,65] According to Henneman, the excitability or threshold level at which a motor unit is recruited is inversely related to the diameter of the motoneuron. Thus the participation of a motor unit in graded motor activity is dictated by the size of its neuron. It appears that slow fibers are innervated by small, low threshold, slow conducting motor nerves. Fast fibers are innervated by larger, higher threshold, faster conducting motor nerves. Thus, slow fibers are recruited first, followed by fast fibers. Studies conducted by other researchers have supported this finding.[3,18,30,49,50,61] Rate coding allows force regulation through summation of the force developed by single twitches. There is a frequency of stimulation above which twitch responses become fused and fibers generated their maximal force. Below the fusion frequency, fibers generate submaximal forces which vary relative to the stimulation frequency.[18,67]

Effects of Muscle-Tendon Architecture

At the level of the gross muscle, the physiological cross-sectional area (PCSA) is most commonly used to indicate a muscle's strength, fiber length, orientation, and type to indicate its maximum velocity of shortening.[95,117] PCSA is calculated by taking the product of muscle mass and the cosine of the pinnation angle, and dividing by the product of fiber length and muscle density. It is important to note that mass alone does not dictate strength, but rather mass and fiber length do so. A muscle with short fibers oriented at some angle relative to the axis of the muscle-tendon complex will generate greater maximum force than a muscle of similar mass that has longer and fewer fibers. Because muscle fibers are composed of serial arrangements of sarcomeres, fiber length affects shortening velocity. Longer fibers have faster shortening velocities, provided the fiber types are similar.

Tendon length and compliance affect muscle-tendon performance.[1,44,45,101,122] A long compliant tendon protects a muscle from injury during sudden imposed stretches. It also transmits muscle force slowly. Short, rigid tendons transmit force rapidly, but provide little protection to the muscle and little potential for storage of elastic strain energy.

Effects of Muscle-Tendon Kinematics

Considerable evidence has been compiled over the years indicating that the amount of force that a muscle can produce depends on its length.[10,21,22,29,52,57,102] Specifically, the force is proportional to the overlap of thick and thin filaments. The fiber length determines the amount of thick and thin filament overlap which determines the number of cross-bridges capable of attaching and developing force. There is an optimal range of muscle fiber length over which the fiber can produce its greatest force. This range occurs at fiber lengths causing the thick and thin filaments to overlap such that all cross-bridges may be active, without overlap of actin filaments from adjacent sarcomeres. At longer fiber lengths not all cross-bridges may contribute to force generation and the force declines. At shorter lengths actin filaments from adjacent sarcomeres begin to interfere with each other and the force also declines. Muscle can also generate passive force. In general, passive force increases gradually from 100 to 130% of rest length and stiffens with increased length. At rest length up to 150%, the deformation is reversible, after which it becomes plastic. The passive properties of muscle may be due to the large molecule titin and membrane structures.

Muscle velocity also affects the force developed. It has been shown that as muscle force increases, the rate of muscle shortening decreases in a hyperbolic fashion.[69,71,82] If muscle is stretched it generates a force greater than its isometric force. Unlike the force-length relationship, the force-velocity relationship has not yet been explained on a precise anatomical basis.

Effects of Muscle Composition

The type of muscle fiber comprising a gross muscle affects the muscle's performance. As discussed previously, myosin molecules in fast and slow twitch skeletal fibers have different ATPase activities.[59,99,103,105] These differences have been correlated with the different shortening velocities that exist between these fiber types.[11,59,103] There are also differences in the troponin C protein in fast and slow twitch fibers. Only one Ca^{+2} site has to be filled to trigger contraction in slow fibers compared to multiple sites in fast fibers.[99] The extent of the T-system varies among different types of muscle fibers. In mammalian muscles, fast twitch fibers have T-systems that are about twice as extensive as those of slow twitch fibers.[80] This property gives rise to faster relaxation rates in fast twitch fibers. Mitochondrial density varies. Fibers relying on oxidative metabolism have greater numbers of mitochondria compared to fibers relying on anaerobic metabolism. These fiber types have the potential to develop force for greater duration compared to glycolytic fibers.

Effects of Contraction History

The contraction history of a muscle-tendon complex can act to reduce or enhance performance relative to how the complex would perform during a standard isometric or concentric action. Fatigue acts to reduce the force that the entire muscle can generate.[6,15,40,55,60,115] However, the mechanisms of fatigue may vary. Basically, anything that inhibits the normal processes of excitation-contraction and coupling described above may cause fatigue. Some of the possible sites where fatigue may be initiated include the central nervous system, the motor end plates, the cytoplasm if pH changes occur, the membranes, and the contractile proteins.

The term *enhancement* has been used in the literature to describe two different effects: (1) elastic energy storage, and (2) force potentiation, an increased force above that of a similar contraction initiated from rest.[4,5,84,113] The first of these effects is related to muscle-tendon elastic properties. The second effect is less understood. However, for both forms of enhancement, the magnitude of the effect depends on several factors. First, for any enhancement to occur a stretch/shortening cycle (eccentric contraction followed by a concentric contraction) must take place. Other factors of relevance are the time delay between the two contraction modes (referred to as coupling time), stretch velocity, initial muscle length prior to stretch, and the amplitude of stretch.[7,16,17,38,39,58,116] The exact mechanisms responsible for enhancement have not been isolated. Storage of elastic strain energy in the tendon and series elastic components of muscle have been suggested as possible sources of the improved mechanical efficiencies reported during certain activities.[2,4,5,28,35,46,113]

Like elastic strain energy, force potentiation is a complex issue. Force potentiation created by a stretch/shortening cycle may be due in part to greater force developed by each cross-bridge attached. There appears to be an optimal eccentric force or amplitude of stretch, below which the magnitude of the force potentiation increases with increased stretch amplitude, and above which it begins to decrease.[4,5] If cross-bridges are stretched too far, then they break and the increased force is lost.

Effects of an Integrated Multiple Muscle System

Under normal conditions muscle-tendon units do not act in isolation. Muscles are influenced by their own actions, which generate specific feedback signals and the signals generated by other muscles and tissues. A motoneuron pool originates in the anterior horn of the spinal cord. Input to a motoneuron pool comes from afferent impulses sent from peripheral receptors, the Renshaw system, and from higher brain centers. These signals may be transmitted along alpha, gamma, or beta neurons.

Feedback to a muscle comes primarily from muscle spindles, and Golgi tendon organs. A muscle spindle is a fusiform capsule attached at both ends to the muscle fibers and arranged in parallel to the fibers. Inside this capsule 2 to 25 are intrafusal fibers. These fibers can contract like extrafusal fibers, but are distinguished because they have centrally located nuclei. At the end of each fiber bundle are two groups of afferent nerves, Ia and II (Ia nerves are larger). Ia afferent nerves connect directly to the motoneuron pool of the muscle and provide excitatory signal. They also connect disynaptically to antagonist muscles to provide inhibitory signals. Group II afferent nerves connect disynaptically to the original muscle only and provide excitatory signals. Ia and II afferent nerves modify their discharge rates when their endings are elongated either by stretching of the muscle or shortening of spindle fibers. Ia afferent nerves are sensitive to length and rate changes, whereas II afferent nerves are primarily sensitive to small length changes.[14,36]

The Golgi organ is located in the aponeurosis and extends from a tendon into the muscle. It has nerve endings sensitive to force. The Golgi organ has a fusiform shape. It is about 650 microns long and 50 microns in diameter. It is innervated by Ib afferent nerves which can generate an inhibitory effect on muscle and a facilitating effect on antagonist muscles, both through disynaptic connections. Renshaw cells, which reside completely in the anterior horn of the spinal cord, are collateral cells that generate negative feedback to nearby neurons. Their role in motor control is not really known.[14]

Muscle-tendon units within the body attach to bones and generate forces to produce joint torques and movement. Muscle-tendon attachment locations directly affect a muscle's potential for moving a limb and generating torque. A muscle-tendon unit with an attachment site relatively far from the joint center will have a mechanical advantage (or expressed more appropriately, less of a mechanical disadvantage since muscle-tendon units usually have severe mechanical disadvantages relative to the external loads they must oppose) compared to a muscle-tendon unit attaching closer to the joint center. However, the latter muscle will have an advantage over the first muscle in producing joint velocity. Thus, relative to performance, joint strength and speed of movement are dictated by the properties of all muscle-tendon units crossing the joint and the locations of their skeletal attachment sites. The musculoskeletal system has considerable redundancy and numerous muscles can create torques about a given joint. These muscles are activated to produce a given torque based on some control scheme that is not understood and likely varies among people and complexities of tasks. Further, there appear to be differences among people in their abilities to realize the full force generating potentials of their muscles and to coordinate the activation of multiple muscles. These differences translate into differences in gross movement performance. A summary of the functions of various muscle-tendon structures is given in Fig. 6.3.

Summary of the Functions of Various Muscle-Tendon Structures	
Structure	**Function**
I. Whole Muscle-Tendon Unit	I. Generate force to stabilize and/or move limb segments. Absorb energy from external sources to reduce loads to other tissues. Store elastic energy for potential reutilization.
II. Fibers	II. Normal cell functions
A. Nuclei	A. Specify DNA sequence for cell proteins
B. Mitochondria	B. Supply ATP through oxidative phosphorylation
C. Ribosomes	C. Produce cell proteins
D. Motor end plate	D. Nerve-muscle fiber interface, filter inputs
E. Membrane Systems	E. Ion barrier, electrical signal conductor
F. Satellite Cells	F. Generate new fibers after injury
G. Sarcomere	G. Basic contractile element
1. Thick Filament	1. Stationary filament
a. Myosin	a. Force development
1) HMM	1) The cross-bridge
a) S1	a) Binding site for actin, site of ATP hydrolysis
b) S2	b) Support for S1
2) LMM	2) Backbone of myosin
2. Thin Filament	2. Translate along thick filament to allow muscle length change.
a. Actin	a. Contains binding sites for myosin
b. Tropomyosin	b. Controls exposure of myosin-sensitive binding sites on actin.
c. Troponin	c. Controls tropomyosin configuration
1) - I	1) Inhibit actin-myosin binding
2) - C	2) Calcium sensitive receptor, controls Troponin-C action.
3) - T	3) Regulate Troponin-Tropomyosin binding
3. M-line	3. Maintain thick filaments in register
4. Z-line	4. Maintain thin filaments in register
5. Titin	5. Provide series elasticity, possibly regulate length assembly
III. Motor Unit	III. Basic neuromuscular element
IV. Tendon	IV. Transmit muscle force, store elastic energy

FIGURE 6.3   A summary of the functions of various muscle-tendon structures.

6.5 Approaches Used to Study Muscle-Tendon Function

The approaches used to study muscle-tendon function are numerous. The review in this section is not intended to be inclusive, but rather to provide a general overview of the wide variety of techniques that have been employed to study those factors affecting muscle-tendon performance described in the previous section. Specifically, studies of the interaction between muscle mechanics and energy utilization, force and neural input, force and length, force and velocity, general performance and architecture, general performance and muscle composition, general performance and contraction history, and general

Summary of Approaches Used to Study Muscle-Tendon Function	
Muscle-Tendon Function	**Approach Used to Study Function**
Muscle mechanics and energy utilization	- isolated muscle preps, muscle stimulation, ergometers, and calorimeters - isolated muscle preps, muscle stimulation, gas analyzers, conversion from oxygen consumption to chemical energy utilization - same approach as above but applied to intact muscle - isolated muscle preps, ergometer, muscle stimulation, quick freeze techniques and chemical analysis - intact muscle, force or pressure transducer, NMR
Force and ... Rate coding	- electrical simulation of varying frequencies, force transducer - voluntary contractions, force transducer, electrodes for recording frequency of muscle activation
Recruitment	- indwelling electrodes to record single motor unit activity, force transducer, gradual increase in voluntary contraction effort - voluntary effort of varying intensity, muscle biopsies to determine motor units depleted of glycogen
Length	-isolated muscle preps, light microscopy, force transducer - intact muscle, extensometer, goniometer or videography, force transducer or dynamometer
Velocity	- isolated muscle preps, lever systems with adjustable loads or electromagnetic ergometers, optical displacement transducers, stimulators - intact muscle, dynamometers

FIGURE 6.4 A summary of various approaches used to study muscle-tendon function.

performance and multiple muscle interactions are discussed. A summary of the approaches used to study muscle tendon function is given in Fig. 6.4.

Muscle Mechanics and Energy Utilization

A variety of methods have been used to determine the energy utilized by a muscle to generate force under various conditions. One approach used for isolated muscle preparations involves placing the muscle in a calorimeter, attaching one end of the muscle to a force transducer or ergometer, activating the muscle, and recording the chemical energy used by the muscle, the work performed, and the heat liberated.[19,48,69] This is the most precise and accurate method, but it is not very applicable to studying muscle *in vivo*. An alternative approach is an indirect method in which the oxygen consumed by the muscle is recorded. The chemical energy used by the muscle is estimated based on the relationship between ATP synthesis and oxygen utilization. This method has been used to study both isolated muscle preparations and muscles acting *in vivo*.[12,13,32,87,90,111]

Summary of Approaches Used to Study Muscle-Tendon Function (Continued)	
Muscle-Tendon Function	**Approach Used to Study Function**
General Performance and ... Muscle Architecture	- dissection, imaging techniques, force transducers, dynamometers
Tendon Architecture	- mechanical testing systems, extensometers, optical tracking devices
Muscle Composition	- same tests as force-length and force-velocity, combined with tests to identify fiber types
Contraction History Fatigue	- electrical stimulation to differentiate central versus peripheral mechanisms - fura-2 and fluorescence microscopy to determine if stimulus is reaching inner cell - pH probes - caffeine administration to determine if cross-bridge is fatigue site - stiffness measurements to determine if force loss is due to reduction in force/cross-bridge or number of cross-bridges
Enhancement	- same as force-velocity, but comparing results from muscle or muscle groups contracting with and without a stretch-shortening cycle - same as mechanics and energetics, but comparing results from muscle or muscle groups contracting with and without a stretch-shortening cycle
Multiple Muscle System	- buckle force transducer to measure force directly - predict force based on model and inputs from EMG, goniometers or videography - estimate force using an inverse dynamics analysis and input from force plates and videography

FIGURE 6.4   (Continued)

Other approaches have quantified the amount of ATP, inorganic phosphate (Pi), and phosphorylcreatine (PCr) before and after muscle activation. These measurements can be used to determine the chemical energy utilized. In one such approach, an isolated muscle is attached to an ergometer and caused to contract. After the contraction the muscle is immediately frozen and the above quantities measured using chemical techniques.[35,118] In a second approach, nuclear magnetic resonance imaging is used to quantify the concentrations of free ATP, PCr, and Pi.[8,118] This method may be used to study muscle *in vivo*, but the signal intensity is very low and multiple trials and signal averaging techniques are required.

Force and Neural Input

Rate coding and recruitment are neural activation characteristics that can regulate muscle force production. Force transducers, neural stimulators, and recording electrodes are the common instruments used to investigate these neural factors although some chemical techniques have also been employed.[3,37,56,64,66,81,92,100] The effect of rate coding has been investigated by stimulating a muscle at different frequencies via its nerve and recording the force developed. Voluntary contractions have also been performed with recording electrodes used to monitor the stimulation frequency over time. The effects of recruitment and the order of motor unit recruitment have been investigated by placing small electrodes within a muscle and recording the electrical activities of single motor units as a person voluntarily contracts the muscle and generates increasingly greater force. Motor units are activated and deactivated in a specific order.[100] The idea of a rank order of recruitment has been supported in several other studies.[18,49,50,61]

Glycogen depletion studies have also been performed to identify which fiber types are involved in different intensities of muscle activation. In these studies, a person utilizes a muscle to produce a given level of force. A muscle biopsy is taken and those fibers depleted of glycogen are identified and classified. In general, oxidative fibers are recruited first, followed by the glycolytic fibers.

Force and Length

The sliding filament theory of muscle length change was developed from results of phase-contrast and interference microscopy[75,76,78] while the mechanisms responsible for the parabolic force-length relationship were demonstrated using X-ray diffraction and electron microscopy.[77] Results from phase-contrast and interference microscopy indicated that the A-band of a muscle fiber does not change length during muscle length change whereas the I-band does. This led to the proposal that filaments slid past one another during muscle length changes. Electron microscopy later identified the individual filaments and the cross-bridges connecting them. Electron microscopy also revealed that cross-bridges could only move about 100 to 140 Å while the length changes observed in the fiber were on the order of 30% of the original length.

This led to the proposal that cross-bridge cycling must occur and that the cross-bridges act as individual force generators. Support for this idea came with the recording of both force and length changes. It was shown that the greatest force occurred when there was optimal overlap of thick and thin filaments, and that the active force decreased in a linear fashion as the length was increased until the thick and thin filaments no longer overlapped, at which time the active force was zero.

Studies of the force-length behaviors of intact muscles have also been performed. These studies rely on force transducers or dynamometers to quantify muscle force or joint torque. Muscle length changes are recorded using video analysis techniques, extensometers, and/or limb displacement measurements combined with musculoskeletal models.

Force and Velocity

The force-velocity relationship of muscle has been derived based on numerous studies of both isolated and intact muscles.[70,71,82,83,106,112] Isolated muscles were stimulated and allowed to shorten while opposed by different load magnitudes. The resistive loads were created with weights and lever systems or electromagnetic devices. The results demonstrate the hyperbolic decrement in velocity for increased load. The experiments conducted on intact muscle involved joint dynamometers which can control either the joint torque or joint angular velocity. The results from intact muscle do not always match those of isolated muscle, but the general trend of decreased velocity for increased force or torque does apply.[112]

General Performance and Muscle-Tendon Architecture

The architectural arrangement of muscle fibers within a muscle affects the amount of force exerted along the axis of the muscle, and the range of muscle lengths over which the muscle can generate force.[23,52,117] Our understanding of the effects of muscle architecture on muscle performance has come from comparative studies of the force-length and force-velocity profiles of muscles that have different architectures. Muscle models have also been used to investigate architectural effects.[52,53,95,98,122]

Tendon structural properties are generally characterized using a mechanical testing system to stretch the tendon while the force and deformation are recorded.[119] These data have been used to determine the tendon's compliance and energy storing capacity.[1,43,44,101]

General Performance and Muscle Composition

The relative compositions of fiber types comprising a muscle affect the muscle's maximum shortening velocity, rate of force development, relaxation rate, fatigue resistance, rate of energy utilization, and power output.[47] Studies illustrating this fact have involved both isolated muscles and intact muscles.[24,31,85,86,111,112] Isolated muscle studies were done by attaching a homogeneous muscle or muscle fiber to an ergometer and recording the force time profile following stimulation. Following the mechanical testing, the muscle was examined via one of the techniques discussed previously to classify the fiber type.[20,25] Different fibers were shown to have different rates of force development and relaxation, different maximum shortening velocities, and different fatigue resistance properties.

Studies of intact human muscles have relied on muscle biopsies to quantify the relative percentage of each fiber type within a muscle combined with joint testing to quantify the torque and power produced by that muscle, and the muscle's fatigue resistance. Testing is usually performed using a single joint and a joint dynamometer or a specific movement such as cycling.[31,56,112] Differences in the rates of energy utilization have also been demonstrated among fiber types.[85,86,118] The techniques used for this determination are the same as those presented in the section on "Muscle Mechanics and Energy Utilization."

General Performance and Contraction History

The techniques used to isolate the mechanisms responsible for muscle fatigue include electrical stimulation, mechanical stiffness measures, and a variety of chemical methods. If a decrement in force results from some mechanisms outside the muscle, then electrical stimulation can be used to elicit a greater force output. For example, if force output during a maximum isometric contraction declines but can be returned to the initial value through external stimulation to the muscle, then the site of fatigue occurred outside the muscle. The site of fatigue within a muscle is difficult to isolate and probably varies depending on the contractile conditions. Fibers have been injected with fura-2 which binds with calcium and can be tracked using digital imaging fluorescence microscopy. This technique has been used to determine whether the excitation signal is carried into the center of the cell and pH probes have been used to determine whether cellular pH changes occur to cause fatigue.

Caffeine has been used to determine whether fatigue is due to insufficient activation of the contractile proteins. Caffeine has the effects of increasing the release of calcium from the SR, reducing the uptake of calcium by the SR, and increasing the troponin C sensitivity to calcium. Thus, if upon administration of caffeine the force increases, then the site of fatigue does not reside in the contractile proteins. Muscle stiffness measurements have been performed in an attempt to determine whether force decrements are due to a decrease in the number of cross-bridges actually generating force or the actual force per cross-bridge. In practice, combinations of these various techniques are used to isolate the site of muscle fatigue.

Force enhancement has been studied in both isolated and intact muscles.[7,16,17,28,38,39,46,84,113] The instruments employed in both cases are similar to those already discussed. Isolated muscle studies involve neural stimulation and muscle force measurements via use of a force transducer or ergometer. Intact muscle studies involve either isolated joint testing with a dynamometer or the determination of gross movement efficiencies by quantifying oxygen consumption and the mechanical work done using force plates and/or some form of motion analysis system. The degree of muscle force enhancement is determined by comparing muscle force or efficiency between muscle actions with and without a stretching-shortening cycle.

General Performance and Multiple Muscle Systems

Historically, three basic approaches have been utilized to predict muscle force *in vivo*. The first approach is direct and relies on some device such as a buckle force transducer to directly monitor the force developed by the muscle. This approach has been used in animal models and to a very limited extent in humans. The second approach is indirect and relies on measurements of specific muscle parameters (e.g., activation levels, kinematics, and architecture) and a suitable mathematical muscle model to compute the forces in individual muscles.[63] The third approach is also indirect, and involves first solving the inverse dynamics problem to determine intersegmental loads (i.e., forces and moments), then utilizing a musculoskeletal model which predicts the behavior of individual muscles when certain criteria like objectives and cost parameters are specified.[33,34,63,97,122]

The instrumentation utilized to obtain the data needed for these approaches includes force plates, electromyography, accelerometers, buckle force transducers, goniometers, and dynamometers. Unfortunately, all of these approaches have limitations and the results obtained are far from consistent for even the most basic human movements. Clearly, our modeling approaches are crude and likely neglect many factors that are critical to the behaviors of muscle-tendon units *in vivo*.

6.6 Summary

In summary, muscle-tendon units involve complex arrangements and interactions of a variety of macroscopic and microscopic structures. A number of techniques have been utilized to identify these structures. Many of these techniques have inherent limitations which necessitate the use of multiple techniques to confirm structural identification. Thus, our understanding of muscle-tendon structure comes from cross-checking the results of many different types of experiments. The contractile characteristics of a whole muscle depend on both gross muscle architecture and the properties of the fibers comprising the muscle. All vertebrate skeletal muscle fibers are similar in their structural arrangement of actin and myosin, but have variations in their membrane structures, density of their mitochondria, specific protein isoforms, and possibly myofibril packing density. These differences, at the molecular level, cause differences in fiber contractile characteristics (i.e., fiber force, maximum shortening velocity, and resistance to fatigue).

At the level of the whole muscle, differences exist among muscles in their arrangements of fibers and percentages of each fiber type. Variations in fiber properties and gross muscle structure mean that different muscles have different contractile characteristics and functions. Our understanding of muscle-tendon function, like muscle-tendon structure, has developed from the findings obtained from use of a variety of technological and methodological approaches. These findings are not always consistent and thus multiple approaches are often required to adequately test various theories of muscle-tendon function.

References

1. Abrahams, M., Mechanical behaviour of tendon *in vitro*, *Med. Biol. Eng.*, 5, 433, 1967.
2. Alexander, R.M. and Bennet-Clark, H.C., Storage of elastic strain energy in muscle and other tissues, *Nature*, 265, 114, 1977.
3. Armstrong, R.B. and Laughlin, M.H., Metabolic indicators of fibre recruitment in mammalian muscles during locomotion, *J. Exp. Biol.*, 115, 201, 1985.
4. Asmussen, E. and Bonde-Petersen, E., Storage of elastic energy in skeletal muscles in man, *Acta Physiol. Scand.*, 91, 385, 1974.
5. Asmussen, E. and Bonde-Petersen, E., Apparent efficiency and storage of elastic energy in human muscle during exercise, *Acta Physiol. Scand.*, 92, 537, 1974.
6. Asmussen, E.M., Muscle fatigue, *Med. Sci. Sports*, 11, 313, 1979.
7. Aura, O. and Komi, P.V., Effects of prestretch intensity on mechanical efficiency of positive work and on elastic behavior of skeletal muscle in stretch-shortening cycle exercise, *Int. J. Sports Med.*. 7, 137, 1986.
8. Bagshaw, C.R, *Outline Studies in Biology: Muscle Contraction*, 2nd Ed., Chapman and Hall, New York, 1993.
9. Bastholm, E., *The History of Muscle Physiology: From the Natural Philosophers to Albrecht Von Haller*, Ejnar Munksgaard, Kobenhavn, 1950.
10. Banus, M.G. and Zetlin, A.M. The relation of isometric tension to length in skeletal muscle, *J. Cel. Comp. Physiol.*, 12, 403, 1938.

11. Barany, M., ATPase activity of myosin correlated with speed of muscle shortening, *J. Gen. Physiol.*, 50, 197, 1967.

12. Baskin, R.J., The variation in muscle oxygen consumption with velocity of shortening, *J. Gen. Physiol.*, 181, 270, 1965.

13. Baskin, R.J., The variation in muscle oxygen consumption with load, *J. Physiol.*, 49, 9, 1965.

14. Basmajian, J.V. and DeLuca, C.J., *Muscles Alive: Their Functions Revealed by Electromyography*, 5th ed., Williams and Wilkins, Baltimore, 1985.

15. Bigland-Ritchie, B., Bellemare, F., and Woods J.J., Excitation frequencies and sites of fatigue, in *Human Muscle Power*, Human Kinetics Publishers, Champaign, IL, 1986, 197.

16. Bosco, C. and Komi, P.V., Potentiation of the mechanical behavior of the human skeletal muscle through prestretching, *Acta Physiologica Scandinavia*. 106(4):467-472, 1979.

17. Bosco, C., Viitasalo J.T., Komi, P.V., and Luhtanen, P., Combined effect of elastic energy and myoelectrical potentiation during stretch-shortening cycle exercise, *Acta Physiol. Scand.*, 114, 557, 1982.

18. Broman, H., DeLuca, C.J., and Mambrito, B., Motor unit recruitment and firing rates interact in the control of human muscles, *Brain Res.*, 337, 311, 1985.

19. Bronk, D.W., The energy expended in maintaining a muscular contraction, *J. Physio.*, 63, 306, 1930.

20. Brook, M.H. and Kaiser K.K., Muscle fiber types: how many and what kind? *Arch. Neurol.*, 23, 369, 1970.

21. Buchthal, F. and Lindhard, J., The physiology of striated muscle fibre, *Det Kgl. Danske Videnskabernes Selskab. Biologiske Meddelelser*, Ejnar Munksgaard Copenhagen, 1939, vol. 14.

22. Buchthal, F., The mechanical properties of the single striated muscle fibre at rest and during contraction and their structural interpretation, *Det Kgl. Danske Videnskabernes Selskab. Biologiske Meddelelser*. Ejnar Munksgaard Copenhagen, 1942.

23. Buchthal, F. and Schmalbruch, H., Motor unit of mammalian muscle, *Physiol. Rev.*, 60, 90, 1980.

24. Burke, R.E., Levine, D.N., and Zajac, F.E. Mammalian motor units: physiological-histochemical correlation of three types in cat gastrocnemius, *Science*, 174, 709, 1971.

25. Burke, R.E., Levine, D.N., Tsairis, P., and Zajac, F.E., Physiological types and histochemical profiles in motor units of the cat gastrocnemius. *J. Physiol.*, 234 723, 1973.

26. Caplan, A., Carlson, B., Fischman, D., Faulkner, J., and Garrett, W., Skeletal muscle, in *Injury and Repair of the Musculoskeletal Soft Tissues*. Woo, S.L.-Y. and Buckwalter, J.A., Eds., American Academy of Orthopaedic Surgeons, Park Ridge, IL, 1988.

27. Catterall, W.A., Excitation-contraction coupling in vertebrate skeletal muscle: a tale of two calcium channels, *Cell*, 64, 871, 1991.

28. Cavagna, G.A., Storage and utilization of elastic energy in skeletal muscle, *Exercise Sports Sci. Rev.*, 5, 89, 1977.

29. Chapman, A.E., The mechanical properties of human muscle, *Exercise Sport Sci. Rev.*, 13, 443, 1985.

30. Clamann, H.P., Gillies, J.D., Skinner, R.D., and Henneman, E., Quantitative measures of output of a mortoneuron pool during monosynaptic reflexes, *J. Neurophysiol.*, 37, 328, 1974.

31. Coyle, E.F., Costill, D.L., and Lesmes, G.R., Leg extension power and muscle fiber composition, *Med. Sci. Sports Exercise*, 11, 12, 1979.

32. Coyle, E.F., Sidossis, L.S., Horowitz, J.F., and Beltz, J.D., Cycling efficiency is related to the percentage of Type I muscle fibers, *Med. Sci. Sports Exercise*, 24, 288, 1992.

33. Crowninshield, R.D., Use of optimization techniques to predict muscle forces. *J. Biomechanical Eng.*, 100, 88, 1978.

34. Crowninshield, R.D. and Brand, R.A., A physiologically based criterion of muscle force prediction in locomotion, *J. Biomechanics*, 14, 793, 1982.

35. Curtin, N.A. and Davies, R.E., Very high tension with very little ATP breakdown by active skeletal muscle, *J. Mechanochemistry Cell Motility*, 3, 147, 1975.

36. Dietz, V., Schmidtbleicher, D., and Noth, J., Neuronal mechanisms of human locomotion, *J. Neurophysiol.*, 42, 5, 1979.

37. Edgerton, V.R., Roy, R.R., Gregor, R.J., Hager, C.L., and Wickiewicz, T., Muscle fiber activation and recruitment, *Biochem. Exercise*, 13, 31, 1983.

38. Edman, K.A.P., Elzinga, G., and Noble, M.I.M., Enhancement of mechanical performance by stretch during tetanic contractions of vertebrate skeletal muscle fibres, *J. Physiol.*, 281, 139, 1978.

39. Edman, K.A.P., Elzinga, G., and Noble, M.I.M., Residual force enhancement after stretch of contracting frog single muscle fibers, *J. Gen. Physiol.*, 80, 769, 1982.

40. Edwards, R.H.T., Human muscle function and fatigue, in *Human Muscle Fatigue: Physiological Mechanisms*, Ciba Foundation Symposium 82, Pitman Medical, London, 1981, 1.

41. Eisenberg, B.R., Quantitative ultrastructure of mammalian skeletal muscle, in *Handbook of Physiology*, Peachey, L.D., Ed., American Physiological Society, Bethesda, MD, 1983, 73.

42. Eisenberg, B.R., Adaptability of ultrastructure in the mammalian muscle, *J. Exp. Biol.*, 115, 55, 1985.

43. Elliot, D.H., Structure and function of mammalian tendon, *Biol. Rev.*, 40, 392, 1965.

44. Elliot, D.H. and Crawford, G.N.C., The thickness and collagen content of tendon relative to the strength and cross-sectional area of muscle, *Proc. R. Soc. London*, 162, 137, 1965.

45. Ettema, G.J.C. and Huijing, P.A., Properties of the tendinous structures and series elastic component of EDL muscle-tendon complex of the rat, *J. Biomechanics*, 22, 1209, 1989.

46. Faraggiana, H.T. and Margaria, R., Utilization of muscle elasticity in exercise, *J. Appl. Physiol.*, 32, 491, 1972.

47. Faulkner, J.A., Claflin, D.R., and McCully, K.K., Power output of fast and slow fibers from human skeletal muscles, in *Human Muscle Power*, Human Kinetics Publishers, Champaign, IL, 1986, 81.

48. Fenn, W.O., The relationship between work performed and the energy liberated in muscular contraction, *J. Physiol.*, 58, 373, 1924.

49. Freund, H.J., Budingen, H.J., and Dietz, V., Activity of single motor units from human forearm muscles during voluntary isometric contractions, *J. Neurophysiol.*, 38, 933, 1975.

50. Freund, H.J., Motor unit and muscle activity in voluntary motor control, *Physiol. Rev.*, 63, 387, 1983.

51. Fung, Y.C., *Biomechanics: Mechanical Properties of Living Tissues*, Springer-Verlag, New York, 1981.

52. Gans, C., Fiber architecture and muscle function, *Exercise Sports Sci. Rev.*, 10, 160, 1982.

53. Gareis, H., Solomonow, M., Baratta, R., Best, R., and D'Ambrosia, R., The isometric length-force models of nine different skeletal muscles, *J. Biomechanics*, 25, 903, 1992.

54. Garrett, W.E. and Best, T.M., Anatomy, physiology, and mechanics of skeletal muscle, in *Orthopaedic Basic Science*, Simon, S.R., Ed., American Academy of Orthopaedic Surgeons, Park Ridge, IL, 1994.

55. Gibson, H. and Edwards, R.H.T., Muscular exercise and fatigue, *Sports Med.*, 2, 120, 1985.

56. Gollnick, P.D., Piehl, K., and Saltin, B., Selective glycogen depletion pattern in human muscle fibres after exercise of varying intensity and at varying pedaling rates, *J. Physiol.*, 241, 45, 1974.

57. Gordon, A.M., Huxley, A.F., and Julian, F.J., The variation in isometric tension with sarcomere length in vertebrate muscle fibres, *J. Physiol.*, 184, 170, 1966.

58. Goubel, F., Muscle mechanics fundamental concepts in stretch-shortening cycle, *Med. Sports Sci.*, 26, 24, 1987.

59. Greaser, M.L., Moss, R.L., and Reiser, P.J., Variations in contactile properties of rabbit single muscle fibres in relation to troponin T isoforms and myosin light chains, *J. Physiol.*, 406, 85, 1988.

60. Green, H.J., Muscle power: fibre type recruitment, metabolism and fatigue, in *Human Muscle Power*, Human Kinetics Publishers, Champaign, IL, 1986, 65.

61. Grimby, L., Motor unit recruitment during normal locomotion. *Med. Sports Sci.*, 26, 142, 1987.

62. Hannerz, J., Discharge properties of motor units in relation to recruitment order in voluntary contraction, *Acta Physiol. Scand.*, 91, 374, 1974.

63. Hatze, H., *Myocybernetic Control Models of Skeletal Muscle*, University of South Africa, Pretoria, 1981.

64. Henneman, E., Somjen, G., and Carpenter, D.O., Functional significance of cell size in spinal motoneurons, *J. Neurophysiol.*. 28, 560, 1965.

65. Henneman, E., Clamann, H.P., Gillies, J.D., and Skinner, R.D., Rank order of motoneurons within a pool: Law of combination, *J. Neurophysiol.*, 37, 1338, 1974.

66. Henneman, E., The size-principle: A deterministic output emerges from a set of probabilistic connections. *J. Exp. Biol.*, 115, 105, 1985.

67. Hennig, R. and Lomo, T., Gradation of force output in normal fast and slow muscles of the rat, *Acta Physiol. Scand.*, 130, 133, 1987.

68. Herman, B. and Lemasters, J.L., *Optical Microscopy: Emerging Methods and Applications*, Academic Press, San Diego, 1993.

69. Hill, A.V., Energy liberation and "viscosity" in muscle, *J. Physiol.*, 93, 4, 1938.

70. Hill, A.V., The variation in total heat production in a twitch with velocity of shortening, *Proc. R. Soc. London*, 159, 596, 1964

71. Hill, A.V., *First and Last Experiments in Muscle Mechanics*, Cambridge University Press, London, 1970.

72. Hille, B., *Ionic Channels of Excitable Membranes*, 2nd Ed., Sinauer Press, Sunderland, MA, 1992.

73. Hochachka, P.W., Fuels and pathways as designed systems for support of muscle work, *J. Exp. Biol.*, 115, 149, 1985.

74. Hochachka, P.W., *Muscles as Molecular and Metabolic Machines*, CRC Press, Ann Arbor, MI, 1994.

75. Huxley, A.F. and Niedergerke, R., Interference microscopy of living muscle fibres, *Nature*, 173, 971, 1954.

76. Huxley, H.E. and Hanson, J., Changes in the cross-striations of muscle during contraction and stretch and their structural interpretation, *Nature*, 173, 973, 1954.

77. Huxley, H.E., The mechanisms of muscular contraction recent structural studies suggest a revealing model of cross-bridge action at variable filament spacing, *Science*, 164, 1356, 1969.

78. Huxley, H.E., *Reflections on Muscle*, Princeton University Press, Princeton, NJ, 1980.

79. Huxley, H.E., The cross bridge mechanism of muscular contraction and its implications, *J. Exp. Biol.*, 115, 17, 1985.

80. Ishikawa, H., Fine structure of skeletal muscle, *Cell and Muscle Motility*, 4, 1, 1983.

81. Kanosue, K., Yoshida, M., Akazawa, K., and Fujii, K., The number of active motor units and their firing rates in voluntary contraction of human brachialis muscle, *Japanese J. Physiol.*, 29, 427, 1979.

82. Katz, B., The relation between force and speed in muscular contraction, *J. Physiol.*, 96, 64, 1939.

83. Komi, P.V., Measurement of the force-velocity relationship in human muscle under concentric and eccentric contractions, in *Biomechanics III*, 3rd International Seminar, Rome, S. Karger, Basel, 1973, 224.

84. Komi, P.V., The stretch-shortening cycle and human power output, in *Human Muscle Power*, Human Kinetics Publishers, Champaign, IL, 1986, 27.

85. Kushmerick, M.J., Patterns in mammalian muscle energetics, *J. Exp. Biol.*, 115, 165, 1985.

86. Kushmerick, M.J., Pattern of chemical energetics in fast- and slow-twitch mammalian muscles, *Biochem. Exercise*, 13, 51, 1983.

87. Kyröläinen, H., Komi, P.V., Oksanen, P., Hakkinen, K., Cheng, S., and Kim, D.H., Mechanical efficiency of locomotion in females during different kinds of muscle action, *Eur. J. Appl. Physiol.*, 61, 446, 1990.

88. Lieber, R.L., *Skeletal Muscle Structure and Function: Implications for Rehabilitation and Sports Medicine*, Williams and Wilkins, Baltimore, 1992.

89. Lexell, J., Henriksson-Larsen, K., and Sjostrom, M., Distribution of different fiber types in human skeletal muscle, *Acta Physiol. Scand.*, 117, 115, 1983.

90. Margaria, R., Positive and negative work performance and their efficiencies in human locomotion, *Int. Z. Angew Physiol. Einschl. Arbeitsphysiol.*, 25, 339, 1968.

91. McMahon, T.A., *Muscles, Reflexes, and Locomotion*, Princeton University Press, Princeton, NJ, 1984.

92. Milnar-Brown, H.S., Stein, R.B., and Yemm, R., Changes in firing rates of human motor units during linearly changing voluntary contractions, *J. Physiol.*, 230, 371, 1973.

93. Nemeth, P.M. and Pette, D., The limited correlation of myosin-based and metabolism-based classifications of skeletal muscle fibers, *J. Histochem. Cytochem.*, 29, 89, 1981.

94. Ogilvie, R.W. and Feeback, D.L., A metachromatic dye-ATPase method for the simultaneous identification of skeletal muscle fiber types I, IIA, IIB, and IIC, *Stain Technol.*, 65, 231, 1990.

95. Otten, E., Concepts and models of functional architecture in skeletal muscle, *Exercise Sports Sci. Rev.*, 16, 89, 1988.

96. Peachey, L.E., Excitation-contraction coupling: the link between the surface and the interior of a muscle cell, *J. Exp. Biol.*, 115, 91, 1985.

97. Pedotti, A., Krishnan, V.V., and Stanley, L., Optimization of muscle-force sequencing in human locomotion, *Math. Biosci.*, 38, 57, 1978.

98. Perrine, J.J. and Edgerton, V.R., Muscle force-velocity and power-velocity relationships under isokinetic loading, *Med. Sci. Sports*, 10, 159, 1978.

99. Perry, S.V., Properties of the muscle proteins: a comparative approach, *J. Exp. Biol.*, 115, 31, 1985.

100. Person, R.S. and Kudina, L.P., Discharge frequency and discharge pattern of human motor units during voluntary contraction of muscle, *Electroencephalograp. Clin. Neurophysiol.*, 32, 471, 1972.

101. Rack, P.M.H. and Westbury, D.R., Elastic properties of the cat soleus tendon and their functional importance, *J. Physiol.*, 347, 495, 1984.

102. Ramsey, R.W. and Street, S.F., The isometric length-tension diagram of isolated skeletal muscle fibers of the frog, *J. Cell. Comp. Physiol.*, 15, 11, 1940.

103. Reiser, P.J., Moss, R.L., Giulian, G.G., and Greaser, M.L., Shortening velocity in single fibers from adult rabbit soleus muscles is correlated with myosin heavy chain composition, *J. Biol. Chem.*, 260, 9077, 1985.

104. Rochow, T.G. and Rochow, E. G., *An Introduction to Microscopy by Means of Light, Electrons, X-Rays, or Ultrasound*, Plenum Press, New York, 1978.

105. Saltin, B. and Gollnick, P.D., Skeletal muscle adaptability: significance for metabolism and performance, in *Handbook of Physiology: Skeletal Muscle*, American Physiological Society, Bethesda, MD, 1983, chap. 19.

106. Spector, S.A., Gardiner, P.F., Zernicke, R.F., Roy, R.R., and Edgerton, V.R., Muscle architecture and force-velocity characteristics of cat soleus and medial gastrocnemius: implications for motor control, *J. Neurophysiol.*, 44, 951, 1980.

107. Spurway, N., Interrelationship between myosin-based and metabolism-based classifications of skeletal muscle fibers, *J. Histochem. Cytochem.*, 29, 87, 1981.

108. Squire, J., *The Structural Basis of Muscular Contraction*, Plenum Press, New York, 1981.

109. Squire, J., *Muscle: Design, Diversity, and Disease*, Benjamin/Cummings Publishing, Menlo Park, CA, 1986.

110. Squire, J., *Molecular Mechanisms in Muscular Contraction*, MacMillan Press, London, 1990.

111. Suzuki, Y., Mechanical efficiency of fast- and slow-twitch muscle fibers in man during cycling, *J. Appl. Physiol.*, 47, 263, 1979.

112. Thorstensson, A., Grimby, G., and Karlsson, J., Force-velocity relations and fiber composition in human knee extensor muscles, *J. Appl. Physiol.*, 40, 12, 1976.

113. Thys, H., Faraggiana, T., and Margaria, R., Utilization of muscle elasticity in exercise, *J. Appl. Physiol.*, 32, 491, 1972.

114. White, D.C.S., *Biological Physics*, Chapman and Hall, London, 1974.

115. Wilkie, D.R., Shortage of chemical fuel as a cause of fatigue: studies by nuclear magnetic resonance and bicycle ergometry, in *Human Muscle Fatigue: Physiological Mechanisms*, Ciba Foundation Symposium 82, Pitman Medical, London, 1981, 102.

116. Wilson, G.J., Elliot, B.C., and Wood, G.A., The effect on performance of imposing a delay during a stretch-shorten cycle movement, *Med. Sci. Sports Exercise*, 23, 364, 1991.

117. Woittiez, R.D., Huijing, P.A., Boom, H.B.K., and Rozendal, R.H., A three-dimensional muscle model: A quantified relation between form and function of skeletal muscles, *J. Morphol.*, 182, 95, 1984.

118. Woledge, R.C., Curtin, N.A., and Homsher, E., *Energetic Aspects of Muscle Contraction*. Academic Press, New York, 1985.

119. Woo, S.L.-Y., Mechanical properties of tendons and ligaments I. Quasi-static and nonlinear viscoelastic properties, *Biorheology*, 19, 385, 1982.

120. Woo, S.Y.-L., An, K., Arnoczky, S.P., Wayne, J.S., Fithian, D.C., and Myers, B.S., Anatomy, biology, and biomechanics of tendon, ligament, and meniscus, in *Orthopaedic Basic Science*, Simon, S.R., Ed., American Academy of Orthopaedic Surgeons, Park Ridge, IL, 1994, chap. 2.

121. Yamaguchi, G.T., Sawa, A.G.U., Moran, D.W., Fessler, M.J., and Winters, J.M., A survey of human musculotendon actuator parameters, in *Multiple Muscle Systems*, Winters, J.M. and Woo, S., Eds., Springer-Verlag, New York, 1990.

122. Zajac, F.E., Muscle and tendon: properties, models, scaling, and application to biomechanics and motor control, *Critical Reviews in Biomedical Engineering*, Bourne, J.R., Ed., CRC Press, Boca Raton, FL, 1989.

7

A Technique for the Measurement of Tension in Small Ligaments

Chimba Mkandawire
Harborview Medical Center

Phyllis Kristal
Harborview Medical Center

Allan F. Tencer
Harborview Medical Center

7.1 Introduction

The goal of this chapter is to present a method for the measurement of the *in situ* tensile force in small ligaments, the ligament tension transducer (LTT), and demonstrate its utility by displaying an application to measuring the properties of the ligaments of the volar side of the wrist. This method allows the static *in situ* force within the bulk of a ligament to be determined without disturbing its functional performance. Before presenting the technique, the significance and history of the study of the biomechanical properties of ligaments will be reviewed; the general mechanical properties of ligaments will be presented since these properties affect the methods by which measurements are made; and the advantages and short-comings of other techniques will be discussed. The LTT technique, its performance and limitations, and an example application will then be covered.

The significance of studying the biomechanical properties of ligaments stems from the benefits provided. Such studies have increased our understanding of ligament behavior, helped to identify key ligaments requiring restoration after injury, and have assisted in identifying materials and tissues with appropriate characteristics that can be used as replacements. For instance, classic studies of the properties of the anterior cruciate ligament of the knee and various materials used for its replacement after injury have allowed selection of materials with appropriate strength and stiffness characteristics. This has led to a high success rate for this common procedure.[24]

7.2 Background

A Short Summary of Experimental Techniques in Ligament Biomechanics

Knee ligament studies have dominated the literature, probably because trauma to the knee is very painful and disabling, yet common, and the ligamentous structures are large and easily identified.[29] Since the majority of knee loads are supported by four ligaments, any ligament tear is functionally disabling due to increasing joint instability. In 1974, Warren et al.[33] published an *in vitro* study of knee medial collateral ligament biomechanics using a radiographic technique to determine ligament strain during functional positioning. In 1975, Noyes et al.[25] described the biomechanics of the ACL of the rhesus monkey. Their study correlated tensile force with strain rate, using isolated bone-ligament-bone preparations mounted to a materials testing machine.

During the 1980s, new transducer techniques that emerged allowed a shift from *in vitro* to *in situ* testing. In 1982, Lewis[22] described an *in situ* study on the human cadaver knee anterior cruciate ligament (ACL) using a buckle transducer to measure tensile force. In 1983, Stone et al.[30] performed an *in vitro* study on the human ACL and an *in vivo* study of the canine ACL using the liquid metal strain gage (LMSG) to measure strain in biological tissue. Also in 1983, Arms et al.[4] published an *in situ* study of the human cadaver medial collateral ligament (MCL) of the knee using a Hall effect transducer. The buckle transducer measures force while the LMSG and Hall effect transducers measure strain and will be described in detail later in the discussion. These devices not only allowed the measurement of *in situ* force and strain, but also could be applied to smaller ligaments. The emergence of the buckle transducer pushed biomechanical ligament analysis ahead in understanding ligament function by directly measuring the tensile force carried by the ligament; unfortunately, application of the transducer prestresses the ligament and changes its operating range.[6,22]

In the late 1980s and early 1990s, studies of ligament function were expanded to the ankle and wrist, with the adaptation of instrumentation used in the large ligaments of the knee to the smaller ligaments of these other joints. In 1988, Renstrom and Arms[26] used the Hall effect transducer to measure *in situ* strain in cadaver ankle ligaments. In 1990, de Lange et al.[15] performed an *in situ* study of the strain in a number of ligaments of the human cadaver wrist. Their group used a biplanar radiography method by which the three-dimensional positions of tantalum balls placed within the ligamentous substance were determined during functional loading of cadaveric wrists. This method produced a large amount of detailed information since strain within different regions of the ligament could be detected. In 1993, Acosta et al.,[1] adapted a smaller Hall effect transducer for use in measuring the *in situ* strains of wrist ligaments. In 1994, Kristal et al.[21] and Weaver et al.[34] developed the ligament tension transducer (LTT) for application to measuring the functional strains in eight ligaments of the volar side of the wrist. This device allowed the study of very small ligaments, less than 1 cm in length, and provided an *in situ* static measurement of force that did not change the function of the ligament. All other techniques described measured ligament strain and only provided indirect indications of mechanical function. On the other hand, the other transducers allow continuous measurement so that dynamic testing can be performed.

Apart from the need to further study smaller ligaments experimentally, mathematical models can be used to describe ligament properties. At the macromolecular level, both tendons and ligaments are primarily made of type I collagen. Considerable attention has been paid to models of tendon mechanical function, but there has been little focus on ligaments. If the cross-sectional shape of a ligament varies during loading, changes in the overall material and mechanical properties occur.[10,36] Ligaments have many different cross-sectional shapes and thicknesses which makes modeling challenging and indicates that experimental measurement will continue to provide a significant source of information.

Comparison of *In Situ* and *In Vitro* Models

In vitro and *in situ* models have been used to evaluate the properties of ligaments. An *in situ* measurement is taken on a ligament that has not been removed from its anatomic setting, while an *in vitro* measurement is taken on a ligament that has been harvested. For determining stress-strain behavior, the *in situ* model comes closer to simulating the *in vivo* behavior. When using an *in vitro* approach, measurement of the initial *in situ* ligament length should be made before removal of the ligament. This defines the operating condition of the ligament, for example, its prestress condition. *In vitro* testing must consider the anatomic directions in which the load is applied, which may not necessarily be along the axes of the ligament fibers. Another difference between the two approaches is that ligamentous specimens tested *in vitro* experience end effects from clamping to the mechanical testing machine. Such enforced boundary conditions change local stress fields about the anchor points, and may cause differences in mechanical behavior. Therefore, one can see that an *in situ* experimental model approximates the *in vivo* condition better than the *in vitro* model does.

Biomechanical Properties of Ligaments

Ligaments do not follow the laws of continuum mechanics, so they cannot be modeled as ideal elastic solids.[17] In this section, solid continuum mechanics aspects are discussed since they provide a framework for understanding ligament behavior. Then, ligament viscoelastic or time dependent properties are demonstrated since they, too, have significant effects on measured properties.

An ideal elastic solid can be modeled using Hooke's law, which states that stress is directly proportional to strain and Young's modulus. From the theory of elasticity, any ideal isothermic and isotropic elastic-solid can be three-dimensionally modeled by the following equations.

$$\frac{\partial T_{ij}}{\partial X_j} + \rho b_i = \rho \frac{\partial^2 u_i}{\partial t^2}; \begin{array}{l} i = 1-3 \\ j = 1-3 \end{array} \tag{7.1}$$

$$T_{ij} = \lambda E_{kk} \delta_{ij} + 2\mu E_{ij} \tag{7.2}$$

$$E_{ij} = \frac{1}{2}\left(\frac{\partial u_i}{\partial X_j} + \frac{\partial u_j}{\partial X_i}\right) \tag{7.3}$$

Eq. 7.1 represents three equations of motion which satisfy force equilibrium. The first term represents the sum of traction vectors expressed in three orthonormal directions. The second term is the sum of all body forces acting on an object. The last term is the sum of all the resultant accelerations; ρ is the mass density, T_{ij} is the stress tensor, and u_i is the displacement vector. Eq. 7.2 is Hooke's law, rewritten in indicial notation. The first term is the Cauchy-Green stress tensor. The second term identifies volumetric strain, and the third term identifies shear strain. E_{ij} is the strain tensor; μ and λ are Lamé constants. Eq. 7.3 is the set of geometric compatibility equations. The strain tensor is a function of orthonormal displacements and lengths. Fifteen unknowns are presented in Eqs. 7.1 through 7.3: 6 stresses, 6 strains, and 3 displacements.

Since most ligaments are tested with uniaxial loading, the theory of elasticity can be reduced to Eq. 7.4, where E is Young's modulus, T_{xx} is uniaxial stress and ε_{xx} is uniaxial strain. This is shown experimentally in Fig. 7.1.

$$T_{xx} = E\varepsilon_{xx} \tag{7.4}$$

We cannot accurately model solid biological tissues as Hookean solids. Biosolids differ from Hookean solids because of their nonlinear characteristics, viscoelasticity, and plasticity. Three phenomena define viscoelasticity: hysteresis, creep, and stress relaxation. Hysteresis, shown in Fig. 7.1, occurs when the stress-strain curve shifts during cyclic loading, since less energy is returned during the unloading phase of the test. Creep, as shown in Fig. 7.2A, occurs when a body continues to deform under a constant

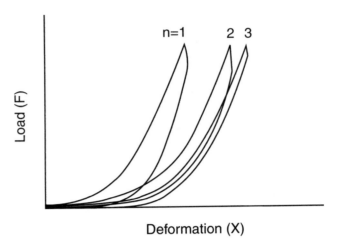

FIGURE 7.1 Example of the results of uniaxial tensile testing of a ligament demonstrating nonlinear response and hysteresis or energy loss with unloading. (*Source*: Fung, Y.C., *Biomechanics: Mechanical Properties of Living Tissues*, Springer-Verlag, NY, 1981. With permission.)

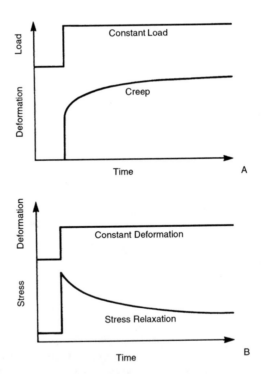

FIGURE 7.2 A. Typical response of a ligament to a step load demonstrating creep or continued deformation. B. Response of a ligament to a step deformation demonstrating stress relaxation. (*Source*: Mow, V.C. and Hayes, W.C., Eds., *Basic Orthopedic Biomechanics*, Raven Press, NY, 1991. With permission from Lippincott Williams and Wilkins.)

stress, preventing the establishment of equilibrium. Stress relaxation, as shown in Fig. 7.2B, occurs when the deformation is maintained constant and the stress decreases.

Fung introduced a mathematical framework to characterize viscoelastic behavior in soft tissues.[18] Fung's law is known as the quasi-linear viscoelasticity law, and shown in Eq. 7.5.

$$\sigma(t) = f\{\varepsilon(t)\} + f'\{\varepsilon(t - \tau); t, \tau\} \tag{7.5}$$

From Fung's law, $\sigma(t)$ and $\varepsilon(t)$ signify stress and strain at any given time t. The term $f\{\varepsilon(t)\}$ represents a function of time-dependent strain, and the $f'\{\varepsilon(t - \tau); t, \tau\}$ term represents a function of the whole time history. Haut and Little have modified this equation in analysis of the biomechanics of rat tail tendon.[19] Later sections of this chapter will present a more in-depth study of ligament viscoelasticity. For more information outside of the scope of this chapter, consult Fung's and Viidiks' chapters in *Handbook of Bioengineering*,[31] and *Biomechanics of Diarthrodial Joints*.[32]

7.3 Measuring Biomechanical Properties of Ligaments *In Situ*

We strongly believe that *in situ* tensile load determination is a more direct measurement of ligament function than *in situ* strain since force must be inferred indirectly from the measured strain.[34] This force estimation may be achieved through the use of Fung's law that was modified by Haut and Little and Butler et al.[11,19] Another indirect method of determining load-carrying capabilities was performed by Huiskes (1991, unpublished). After ligament strain was measured *in situ*, each ligament biomechanical unit was removed from the wrist and force-displacement curves were measured *in vitro*. A direct *in situ* tensile force measurement technique eliminates the potential uncertainties associated with measuring *in situ* strain and then converting the data to force. In the next section, the methodology, strengths, and weaknesses of the measurement techniques for *in situ* strain and force measurement in ligaments are discussed. The description of the ligament tension transducer concludes this chapter.

Liquid Metal Strain Gage

The liquid metal strain gage (LMSG) transducer system is the combination of an LMSG as the primary sensing element and its supporting electronic hardware. The LMSG is an electromechanical transducer; it reads a length change and outputs a voltage. The LMSG is a mercury-filled silastic tube incorporated into electrical wire. This simple configuration is a powerful feature because mercury is a naturally occurring liquid-element that is very conductive and the system is highly compliant while accommodating large strains. The LMSG has a linear response when the operating range is kept below 40% strain, due to direct extension of the length of the silastic tube and a corresponding decrease in tube cross-sectional area, both of which change resistance across the gage (Fig. 7.3). If an LMSG is stretched above 140% of its total length, the electrical response deviates into nonlinearity. The LMSG can be used *in vivo*, to measure strain history in dynamic loading. Brown reported the LMSG dynamic response to be flat to 50 Hz and without phase shift.[9]

The electrical resistance of the LMSG changes with the change in length and cross-sectional area of the mercury column within; therefore, the voltage drop across the supporting electronic hardware will correlate to a specific length change. Stone et al.[30] have derived this relationship, which is shown in Eq. 7.6, where R is the resistance of the gage, ΔR is the change in resistance due to strain, and ε_l is the axial strain along the length of the gage.

$$\frac{\Delta R}{R} = 2\varepsilon_l \tag{7.6}$$

The LMSG can be used in either of two testing configurations: a Wheatstone bridge or a series circuit.[23] In the Wheatstone bridge, the LMSG is placed in series with one arm of the bridge. The series-circuit configuration has the LMSG in series with a drop-down resistor. The outputs of both circuits must be amplified to increase resolution. Meglan[23] pointed out that the series-circuit configuration is ten times more sensitive than the Wheatstone bridge; however, the output of the series circuit is not truly linear. The Wheatstone bridge has great linear response, but lacks sensitivity. The sensitivity of the series-circuit configuration can be enhanced by increasing the current passing through the system, but that would

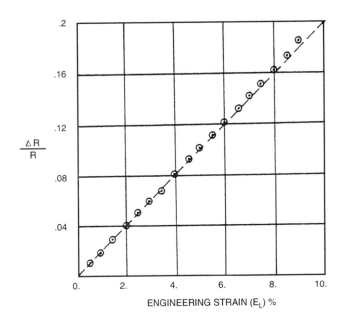

FIGURE 7.3   Liquid metal strain gage (LMSG) performance in terms of change in resistance divided by original resistance with increasing engineering strain. (*Source*: Stone, J.E., Madsen, N.H., Milton, J.L., Swinson, W.F., and Turner, J.L., *Experimental Mech.*, 132, June 1983. With permission from Sage Publ.)

increase heating of the LMSG. Too much heat generation causes LMSG response to become more nonlinear. Another caveat of the series-circuit configuration is that the value of the drop-down resistor has to be chosen carefully. The purpose of this resistor is to minimize the effects of electrical heating.

The LMSG has been used on cadaver knees *in situ* for quasi-static tests[23,30] and to measure strains in cruciate ligaments. Gages can be attached to these ligaments by either of two methods. One method is to use a contact cement that bonds well to biological tissue. The lead wires should be secured to the tibia and femur at the ligament insertion points. This action assures parallel alignment of the gage with respect to the ligament. The second method of gage attachment is to suture the lead wires of the gage to the ligament itself. The LMSG should be pre-stressed when attached to the ligament, so it is operating in its linear range. Strain of 5 to 10% is ideal, assuming no compression will take place. This is an important precaution because the LMSG can only measure tensile strains.

The LMSG has several limitations. A great deal of care must be exercised when handling an LMSG. If the silastic tubing is over-stretched, it may rupture and leak mercury into the environment. A typical LMSG has a shelf life of 6 months, because the mercury slowly oxidizes out of the silastic tubing. The anchoring method of the LMSG is not completely reliable. The suture method requires less space to mount the gage, but the ligament must be pierced for anchoring. The act of piercing holes into the ligament changes its properties. The suture method allows potential slack in the LMSG-ligament system, introducing hysteresis. The glue method of attachment is fragile. Also, the LMSG requires a minimum amount of space to operate, but cannot be used on small ligaments in confined spaces. The LMSG records surface strain between its attachment points, not necessarily the average strain throughout an axial cross-section.

The positive characteristics of the LMSG outweigh its limitations. The output of the LMSG is very linear when used with a Wheatstone bridge. The linear operating range of the transducer is very large, so it is suitable for biologic tissue response. The LMSG is inexpensive, easy to use, easy to calibrate, fast to set up, and capable of both static and dynamic strain measurement. The LMSG is also easy to manufacture. Brown et al.[9] made their own because the commercially available products were too large for *in vivo* studies. However, it must be emphasized that this device measures ligament strain, not force, which still must be determined indirectly.

Hall Effect Transducer

The Hall effect strain transducer (HEST) is the combination of a Hall effect transducer and supporting electronics. The HEST is an electromagnetic device; it reads a change in a magnetic field and outputs a voltage drop that is proportional to the magnetic field. As shown in Fig. 7.4A, a simple Hall effect transducer is a small instrument made of only three parts: a magnetic wire, a Teflon casing, and a Hall effect semiconductor. The semiconductor is anchored to the Teflon casing and the magnetic wire is free to slide in and out of the casing. The midrange response of an HEST, from 10 to 40% strain, is linear, as shown in Fig. 7.4B, but measuring at its extremes produces very nonlinear results.[4] The HEST is closely related to the linear variable differential transducer (LVDT) in principle. The Hall effect semiconductor detects the proximity of a permanent magnet; consequently, it produces a voltage drop that is proportional to the strength of the magnetic field.[26]

The HEST device requires only a current source and a precision amplifier. Because the operating range is from 10 to 40%, it is extremely important to anchor the HEST with 20% strain onto the ligament in its rest position. Otherwise, one runs the risk of measuring in a nonlinear range with a linear calibration curve. There are two methods of anchoring an HEST to a ligament: suturing, or piercing the ligament with barbs. Both methods anchor the device by piercing the ligament substance.

The HEST is extremely versatile. Acosta et al.[1] have used the HEST in osteoarthritic human cadaver wrists to measure *in situ* strain in the dorsal and palmar distal radio-ulnar ligaments (DRUL) before and after reconstruction of the distal radio-ulnar joint (DRUJ). Cawley et al.[12] have used the HEST to define *in situ* biomechanical parameters of ankle collateral ligaments during physiologic foot motion. The HEST has been used by Arms et al.[4,5] in cadaver knees to define MCL and ACL properties. Erickson et al.[16] studied dynamic *in vitro* properties of human MCL and ACL in prophylactic knee braces using HEST devices applied *in vivo*.

Buckle Transducer

The buckle transducer works by slightly deflecting the normal configuration of a load-carrying flexible element in three-point bending due to interaction with the ligament. Tension in the ligament fibers causes the ligament to straighten, thereby bending the crossbar and frame of the regular buckle transducer and bending the buckle beam of the modified buckle transducer.[6]

The design of the transducer is specific to the individual ligament. The design, illustrated in Fig. 7.5, is based on the following parameters: (1) ligament parameters: the ligament thickness t, the length L_l, and the expected maximum tension T; (2) transducer performance parameters: the tolerable amount of ligament shortening due to transducer implantation, S_t; (3) transducer material parameters: Young's modulus and yield strain of the chosen metal must be known, and (4) transducer geometric parameters: the minimum width of the transducer b, and the transducer length L_i.[6]

$$\sec \theta = \left[S_t \left(L_l / L_i + 1 \right) \right] \tag{7.7}$$

From Eq. 7.7, we see how the transducer offset angle θ is determined from the transducer geometry. The offset angle can be used to calculate the section modulus of the beam, where the maximum strain is set at the beam's midsection. For this calculation, the transducer is modeled as a simply supported beam in bending, affected by an applied load P, as shown in the top portion of Fig. 7.5. The tensile force can be determined from the product of the section modulus and the strain gage output. Eq. 7.8 shows how the dimensions of A and H_c are determined transducer parameters. The d_c is the center of deflection of the transducer, and L_c is the width of the clip.

$$\operatorname{Tan} \Theta = \left[2 \left(A - H_c + t - d_c \right) \right] / (Li - Lc) \tag{7.8}$$

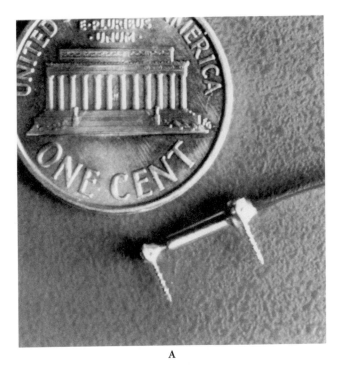

A

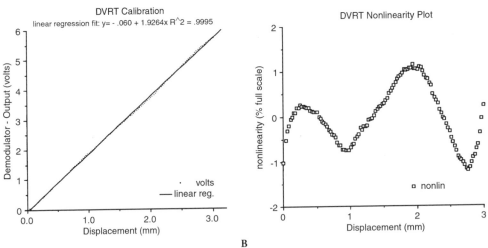

B

FIGURE 7.4   A. Photograph of a Hall effect strain transducer (HEST). B. Ligament strain and resulting force for two different ligaments with and without the buckle transducer indicating the pre-stress effect of the transducer itself. (*Source*: An information brochure, MicroStrain, Burlington, VT. With permission.)

The buckle transducer strain gages form two arms of a Wheatstone bridge. Each strain gage is 120 ohms.[2,3,6] The buckle transducer is attached to a ligament simply by snapping both halves of the transducer together, with the ligament between the halves. During installation, it is important to keep in mind that if too much tissue is inserted, excessive ligament shortening occurs. If not enough ligament tissue is inserted, the signal-to-noise ratio will be too small.[6,22]

Once the transducer is placed on the ligament, the transducer can be calibrated *in situ*. This is done by clamping forceps on the ligament, only a few millimeters from the buckle frame, and then looping a string through the forceps. The other end of the string is attached to a calibrated spring scale. Pulling

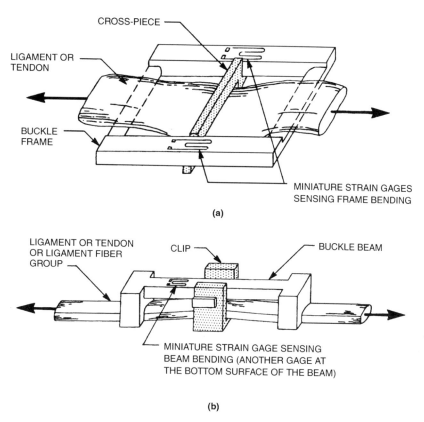

FIGURE 7.5   Schematic diagram of a buckle transducer. (*Source*: Barry, D. and Ahmed, A.M., *J. Biomech. Eng.*, ASME, 108, 149, 1986. With permission.)

on the scale applies a known force through the buckle and ligament, and results in a measurable buckle response.[22]

The effect of a poorly pre-conditioned ligament is more apparent in a buckle transducer than any other device.[6] If the tendon or ligament was not pre-cycled long enough before testing, a noticeable drift in response will be witnessed, as shown in Fig. 7.6. This drift in response is due to the morphological changes of the tissue; moreover, the cross-sectional area changes when the tissue is loaded infrequently, resulting in poor repeatability. The mere act of attaching the buckle transducer onto a ligament causes changes in its length. Once the buckle is locked in place, the resting length of the tissue is shortened because of the path it must take. The presence of the buckle transducer changes the local stresses and boundary conditions at the site to which it is attached.[6] Shortening the ligament changes its stiffness, pre-stressing.[6] Because of these effects imposed on the ligament, it is essential to test the transducer for repeatability during calibration. The main advantage of the buckle transducer is that it measures bulk ligament force directly.

Roentgenstereophotogrammetric Analysis

Stereophotogrammetry is the use of multiple two-dimensional pictures of three-dimensional objects to reassemble a three-dimensional image.[27] The term stereo indicates the reconstruction process of 3-D image building and the prefix *roentgen* indicates that X-rays are used to obtain the image. Roentgenstereophotogrammetry analysis (RSA) is a three-dimensional radiographic technique used to study joint motion pathways. While rigid body joint motion is the primary focus of this technique, it can also be

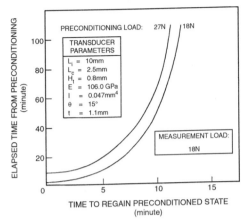
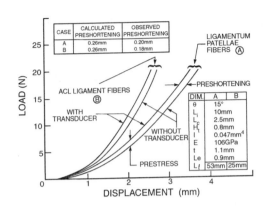

FIGURE 7.6   A. The relationship of the time required for a ligament with a buckle transducer attached to regain its pre-conditioned state based on the time elapsed from pre-conditioning. B. Ligament strain and resulting forces for two different ligaments with and without the buckle transducer indicating the pre-stress effect of the transducer. (*Source*: Barry, D. and and Ahmed, A.M., *J. Biomech. Eng.*, ASME, 108, 149, 1986. With permission.)

used to determine *in situ* strains in soft tissues. Tantalum pellets are used as X-ray markers because of their excellent radiopaque characteristics and biocompatability.[28]

The measurement is performed in two steps (Fig. 7.7). In the first step, after using calibration objects of known shape to locate the two X-ray sources, the intersection between the vectors from the X-ray source to the same point on the X-ray in each of the two planes defines the three-dimensional coordinates of the object to be reconstructed. In the second step, the changes in position of the object after loading can be defined using standard kinematic techniques. For ligament strain measurements the tantalum balls placed into the ligament substance are considered as points and the magnitude of the translation vector divided by its initial (unloaded) magnitude defines the strain of that tissue segment.[14,35]

An experimental setup performed by de Lange et al. is shown in Fig. 7.8. Two roentgen tubes (D) are used to radiograph the specimen. A hand-wrist joint specimen (A) is placed in front of a reference plate (C). Hand movements are controlled by a motion constraint device, and springs (B) are used to load the tendons during testing.[15] RSA has been used successfully in the knee,[7,8,20] wrist,[13-15] and the foot for *in vivo*, *in vitro*, and *in situ* studies.

The successful use of the RSA technique requires accurate knowledge of the locations of the X-ray sources. Therefore, the precision of the calibration process is of fundamental importance. The process is performed on a structure that has known dimensions and is outfitted with tantalum markers; moreover, it is recommended that nine markers which are not coplanar with each other be used.[28] The markers in the test cage function as calibration points, and are X-rayed on the same film as the object. Calibration markers and object markers are exposed from the two separate roentgen foci. The cage markers are of two kinds: fiducial marks and control points. The fiducial marks are used for projective transformations of the image points to the laboratory coordinate system. The control points are used for determining the roentgen foci positions in the same (fiducial) coordinate system. Finally, the three-dimensional coordinates of an object in the test cage can be determined by locating the intersection of the vectors between the roentgen foci and the transformed image points.[27]

This technique has several advantages and disadvantages. The calibration procedure is complex and long. Roentgen film cassettes are not uniformly flat, and that will affect the geometry of the system. It is difficult to maintain specimen alignment throughout an entire range-of-motion recording. The extreme markers must be in the same locations, from one specimen to another.[20] The pellets must be inserted into the ligament by opening a space and gluing the pellets in place.[20] Finally, only static measurements can be made. The system is expensive, and a risk of radiation exposure exists.

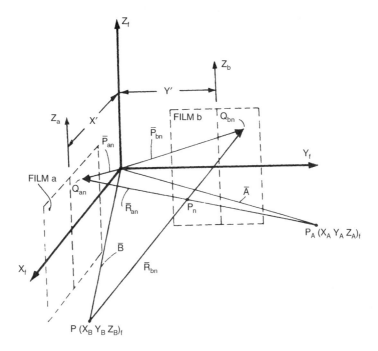

FIGURE 7.7 Schematic diagram of determination of location of a marker point in 3-D space using roentgenstereophotogrammetric analysis (RSA). P and P_A are ideal locations of the X-ray point sources. The vectors Q_{an} and Q_{bn} connect the X-ray sources and the image of the point on each radiograph. P_n is the point in space. (*Source*: Huiskes, R., Kremers, J., Lange, A., de Woltring, H.J., Selvik G., and van Rens, T.J.G., *J. Biomech.*, 18, 559, 1985. With permission from Elsevier Science.)

RSA has two major advantages that all the other transducer systems lack. First, it has been used successfully to make *in vivo* measurements since the placement of tantalum balls into the bones of volunteers has been well tolerated. Second, other techniques only measure bulk tissue strain at the location of the transducer. Arms et al.[4] have shown that the MCL has consistently different strain patterns between the proximal, middle, and distal segments of the anterior and posterior borders. Butler et al. have shown similar findings in the ACL.[11] With RSA, one can measure the local strain wherever two tantalum markers exist. RSA allows the biomechanist to determine complete ligament strain, including bending of the ligament around a bony prominence. Further, RAS has no effect on ligament strain due to application of the technique, unlike the buckle transducer which pre-strains the ligament with insertion.

7.4 Ligament Tension Transducer System

The ligament tension transducer system (LTTS), shown in Fig. 7.9, is based on the qualitative test for ligament integrity performed in surgery which consists of simply pulling the ligament in question in a direction transverse to its long (functional) axis, and estimating its tension. In addition, this method of displacing a cable segment of known length transversely and measuring the transverse force and deformation is used for the quantitative measurement of cable tension in cable rigged structures (such as sailboat masts). In the ligament testing version, a linearly variable differential transformer (LVDT) is used to measure the small transverse deformation applied, and a small load cell provides the force required to do so. During testing, the transducer and specimen must be fixed in space. The probe is placed beneath the ligament being studied, and the displacement screw is turned to first engage and then displace the ligament. The LTTS has been used in two wrist ligament studies. Kristal et al.[21] used it on five ligaments in seven cadaveric hand specimens to determine which ligaments act as key passive motion limiters. An expanded study by Weaver et al.[34] tested eight wrist ligaments to increase the comprehensiveness of the

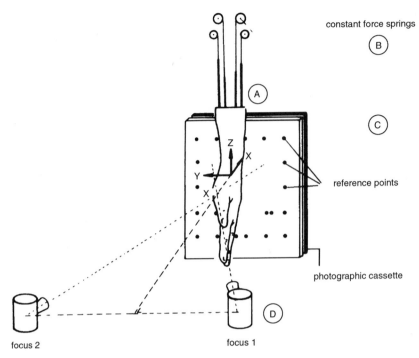

FIGURE 7.8 Example of an experimental setup of the X-ray cartridge, sources, and specimen for RSA. (*Source*: Huiskes, R., Kremers, J., Lange, A., de Woltring, H.J., Selvik G., and van Rens, T.J.G., *J. Biomech.*, 18, 559, 1985. With permission from Elsevier Science.)

experiment. It is important to note that most of the ligaments tested were very small, less than a centimeter in length.

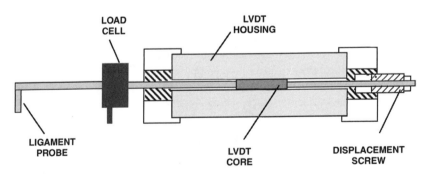

FIGURE 7.9 Schematic diagram of the ligament tension transducer. The probe tip fits behind the ligament. The load cell measures the force required to displace the ligament transversely. The LVDT measures the displacement of the probe which is controlled by the displacement screw. (*Source*: Kristal, P., Tencer, A.F., Trumble, T.E., North, E., and Parvin, D., *J. Biomech. Eng.*, ASME, 115, 218, 1993. With permission.)

The LTTS determines the tensile load in a ligament of known length by measuring the magnitude of the deflection and the force required to do so. For a cable that has a circular cross-sectional area, (Fig. 7.10), the tension in the deformed cable is given in Eq. 7.9.

$$T' = T + K(L' - L) \tag{7.9}$$

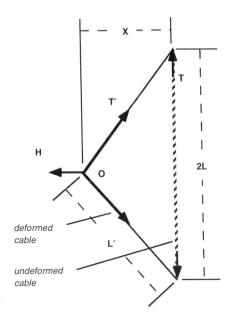

where:
T′ = tension in the deformed cable
T = tension in the undeformed cable
K = stiffness of the cable
L′ = length of the deformed cable
L = length of the undeformed cable
*cable is deformed by pulling it laterally

At point O where the perpendicular
deforming load is applied:

T′ = 1/2 * (L′ / X)* H,

where:
X = imposed perpendicular deformation
H = applied perpendicular force

The tension in the cable is:

T = 1/2 * (L′/X) * H - K (L′ - L)

FIGURE 7.10 Resolution of ligament tensile force using the ligament tension transducer system (LTTS). (*Source*: Weaver, L., Tencer, A.F., and Trumble, T.E., *J. Hand Surg.*, 19A, 464, 1994. With permission from W.B. Saunders.)

where, T' is the tension in the deformed cable, T is the true tension in the undeformed cable, K is the stiffness of the cable, L is the length of the undeformed cable, and L' is the length of the deformed cable. Then,

$$T' = H/2(L'/X) \tag{7.10}$$

Considering equilibrium at the point where the lateral load is applied, Eq. 7.10 states that the sum of the forces is zero. H is the applied lateral load, and X is the imposed lateral deformation. By combining Eqs. 7.9 and 7.10, the tension in the cable can be determined from Eq. 7.11.

$$T = H/2(L'/X) - K(L' - L) \tag{7.11}$$

The first term describes the force balance at the point when H is applied. The second term, the ligament elongation term, describes how the deformed length and stiffness of the cable add to the initial tension in the cable.

The measurement verification process is performed in three steps: verification of the theory using a circular nonbiological cable; *in vitro* comparison of measured to known tension in a typical ligament; and *in situ* ligament tension verification. The test using a circular cross-section cable is necessary to verify the fundamental theory. A nylon cable can be used with a materials testing machine for this step.[21] The load cell of the materials testing machine should be attached to the nylon cable, so recordings of the actual tension in the undeformed cable can be made for comparison to those measured by the LTTS. During this step, it is important to test the effect of nonperpendicular probe orientation.

Bone-ligament-bone preparations should be used for the *in vitro* verification step. Similar to the round cable calibration, the ligament preparations can be placed in a material testing machine, with one end of the ligament attached to the load cell so that the true bulk ligament load is known. Ligaments are more challenging to test than cables for several reasons. Ligaments are not perfectly round, and typically have varying cross-sections along their lengths. Their viscoelastic behavior causes creep when the transverse load is applied. Another problem involves the stiffness term, shown in Eq. 7.11. If it must be included, then the stiffness of the ligament must be determined separately, adding considerable complexity to the measurement procedure, similar to the problem encountered by techniques that measure only strain. A solution is to choose a transverse deformation[21] that makes the stiffness term insignificant. See Table 7.1.

TABLE 7.1 A Summary Comparison of Different Techniques for Measurement of Ligament Strain and Force

Author	Device	Type	Comments
Stone et al.[30]	Liquid metal strain gage	Strain	Static and dynamic measurements of large strains possible
Arms et al.[4]	Hall effect strain transducer	Strain	Static and dynamic measurements, local strains
Barry and Ahmed[6]	Buckle transducer	Force	Static and dynamic measurements, preload ligament
Huiskes[20]	Roentgenstereophotogrammetry	Strain	Static, noncontact so no loading effect, multilocation possible
Kristal[21]	Ligament tension transducer	Force	Static, small ligaments measurable, does not preload ligament

Studies performed by Kristal et al.[21] have shown that during the *in vitro* ligament calibrations, the LTTS was accurate to within 8%. Kristal also pointed out that the LTTS tends to overestimate higher loads and underestimate lower loads. Nonperpendicular probe orientation increases the force required to laterally deform the ligament. An offset of 10° increases the error by 1%; an offset of 20° increases the inaccuracy by 6% (Table 7.2). For the lengths of ligaments encountered in the wrist studies, a transverse displacement of 0.50 mm was imposed so as to neglect the stiffness term in the force calculation. To test for reproducibility, fresh-frozen specimens were thawed, tested, refrozen, thawed and tested again. Thus testing encompassed specimen setup as well as LTTS errors. The overall mean ratio of measured axial tension between first and second trials of any ligament was found to be 1.05 with a standard deviation of 0.29.[21]

TABLE 7.2 Effects of Error in Ligament Length Measurement and Nonperpendicular Probe Alignment on Measured Ligament Tension

Variable	Estimated Error	Error in Measurement
Ligament length	0.30 mm	5.2%
Probe orientation	10° from perpendicular	1.0%
	20° from perpendicular	6.0%

Several assumptions are made when using this technique. One involves estimating the free length of a ligament which may have a broad attachment area. Typically a pair of modified calipers is slid under the ligament until the jaws contact bone. This free length measurement may underestimate the true free length of the ligament. A second assumption is that the bones to which the ligament attaches do not move during the measurement procedure. This can be tested by placing a displacement gage on the bones to which the ligament is connected and determining whether any displacements occur to the connecting bones during the measurement procedure. A third assumption is that the ligaments do not bend around bony prominences. Since some do, which changes the pure tensile force in the ligament to combined tension and bending, the technique cannot be used for these ligaments.

This technique and the others have certain advantages and disadvantages that are summarized in Table 7.1. The LTTS measures tensile force directly. It has minimal effect on the tissue it measures, for example, and does not cause ligament shortening as the buckle transducer does. The LTTS is not anchored to a ligament in the manner that the LMSG and HEST require for operation, so the ligament is not damaged during testing. It is possible for ligament damage to occur when the probe is placed behind a ligament, but this problem can be avoided if the ligament probe tip is bluntly machined. The LTTS measures an average tensile load, unlike the local tensile force measured by a modified buckle transducer or local strains measured by an LMSG or HEST. The LTTS components are moderately inexpensive to manufacture or purchase, and simple to assemble.

As an example of the potential of this technique, a previous experiment on wrist mechanics is briefly described. Five upper extremity specimens were obtained and eight ligaments on the palmar side of the wrist were identified (Fig. 7.11). The specimen was mounted in a positioning frame (Fig. 7.12), which permitted controlled measurable orientation of the hand and provided a stable platform for the ligament tension measurements. At each position of the hand, the ligament tension transducer was oriented perpendicular to the long axis of the ligament to be tested. The probe was hooked behind the ligament and the displacement screw adjusted until the probe tip was not in contact with the back surface of the ligament. It was then moved outward (i.e., transverse to the axis of the ligament) until contact was achieved. From this point, the load applied to laterally deform the ligament was recorded by a load cell (Model 31, Sensotec Precision Miniature Load Cells, Columbus, OH) and displacement by a linearly variable differential transformer (Model 100 DC-D, Shaevitz Engineering, Pennsauken, NJ). The lateral deflection was stopped at 0.5 mm. At that point the magnitude of the load was monitored until it stabilized (i.e., stress relaxation stopped). Since the actual deforming load was very small, that usually occurred within 30 seconds.

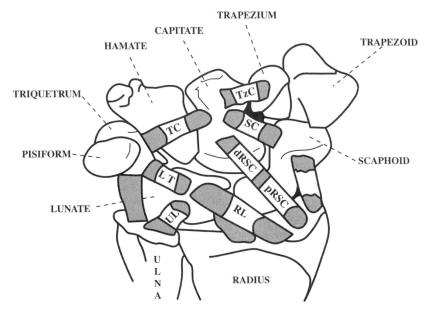

FIGURE 7.11 Ligaments of the palmar side of the wrist tested using the ligament tension transducer system. (*Source*: Weaver, L., Tencer, A.F., and Trumble, T.E., *J. Hand Surg.*, 19A, 464, 1994. With permission from W.B. Saunders.)

After all testing was completed, the minimum lengths between ligament attachment sites were measured using a caliper (Enco Manufacturing Co., Santa Ana, CA) whose knife edge jaws were slid under the ligament and expanded until they encountered the bone ligament junction. This gave the minimum length between bone attachment sites. The error in ligament length from repeated measurements was

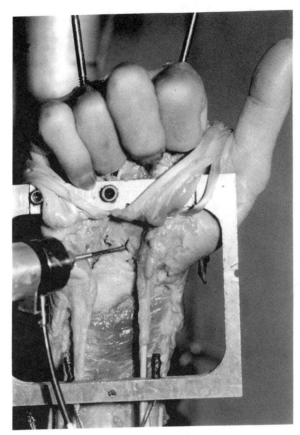

FIGURE 7.12 Experimental setup for measurement of ligament tension using the LTTS. *Source*: Weaver, L., Tencer, A.F., and Trumble, T.E., *J. Hand Surg.*, 19A, 464, 1994. With permission from W.B. Saunders.)

0.30 mm, resulting in a 5.2% error in axial tension. An example of the results is shown in Fig. 7.13. As the hand was moved from radial deviation (the thumb points outward with the palm of the hand facing upward) to ulnar deviation (the thumb points inward and movement from radial to ulnar deviation involves motion in the plane of the palm), the radiolunate and ulnolunate ligaments appear to be the key stabilizers to excessive motion. The radiolunate ligament tension increases with maximum radial deviation while the tension in the ulnolunate ligament is relatively small, and the converse is true as the hand moves to maximum ulnar deviation. The following conclusions were made from the study: (1) the palmar ligaments of the wrist have inherent tension, even in the neutral positioned and unloaded wrist; (2) various ligaments play roles as passive stabilizers at the ends of the ranges of motion of the wrist, and (3) some ligaments have significantly greater tensions than others in any position.

7.5 Summary

A variety of techniques have been developed for measurement of soft tissue functional properties. *In situ* testing causes the least disturbance and should therefore provide the most accurate representation of ligament function. Measurement of strain provides only an indirect measure of the load carrying function of the ligament. Of more benefit is the measurement of ligament load directly. Of two transducers capable of measuring load directly, both the buckle transducer and the ligament tension transducer system (LTTS) have advantages. The buckle transducer can measure dynamic loads in a ligament but its installation pre-stresses the ligament tested. The LTTS can only measure static loads; however, it can be used on very small ligaments (less than 1 cm) and does not pre-stress the ligament.

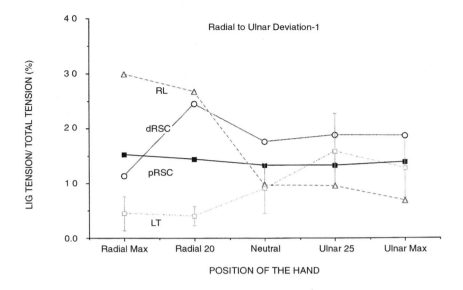

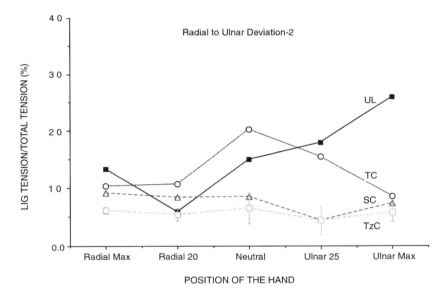

FIGURE 7.13 Example data from testing radial and ulnar deviation. (*Source*: Weaver, L., Tencer, A.F., and Trumble, T.E., *J. Hand Surg.*, 19A, 464, 1994. With permission from W. B. Saunders.)

References

1. Acosta, R., Hnat, W., and Scheker, L.R., Distal radio-ulnar ligament motion during supination and pronation, *J. Hand Surg.*, 18, 502, 1993.
2. Ahmed, A.M., Burke, D.L., Duncan, N.A., and Chan, K.H., Ligament tension pattern in the flexed knee in combined passive anterior translation and axial rotation, *J. Orthop. Res.*, 10, 854, 1992.
3. Ahmed, A.M., Hyder, A., Burke, D.L., and Chan, K.H., *In vitro* ligament tension pattern in the flexed knee in passive loading, *J. Orthop. Res.*, 5, 217, 1987.
4. Arms, S.W., Boyle, J., Johnson, R., and Pope, M., Strain measurement in the medial collateral ligament of the human knee: an autopsy study, *J. Biomechanics*, 16, 491, 1983.

5. Arms, S.W., Pope, M.H., Johnson, R.J., Fischer, R.A., Arvidsson, I., and Eriksson, E., The biomechanics of anterior cruciate ligament rehabilitation and reconstruction, *Am. J. Sports Med.*, 12, 8, 1984.

6. Barry, D. and Ahmed, A.M., Design and performance of a modified buckle transducer for the measurement of ligament tension, *J. Biomechanical Eng.*, 108, 149, 1986.

7. Blankevoort, L., Huiskes. R., and de Lange, A., Helical axes of passive knee joint motions, *J. Biomechanics*, 23, 1219, 1990.

8. Blankevoort, L., Huiskes. R., and de Lange, A., Recruitment of knee joint ligaments, *J. Biomechanical Eng.*, 113, 94, 1991.

9. Brown, T.D., Sigal, L., Njus, G.O., Njus, N.M., Singerman, R.J., and Brand, R.A., Dynamic performance characteristics of the liquid metal strain gage, *J. Biomechanics*, 19, 165, 1986.

10. Butler, D.L., Grood, E.S., Noyes, F.R., Zernicke, R.F., and Brackett, K., Effects of structure and strain measurement technique on the material properties of young human tendons and fascia, *J. Biomechanics*, 17, 579, 1984.

11. Butler, D.L., Sheh, M.Y., Stouffer, D.C., Samaranayake, V.A., and Levy, M.S., Surface strain variation in human patellar tendon and knee cruciate ligaments, *J. Biomechanical Eng.*, 112, 38, 1990.

12. Cawley, P.W. and France, E.P., Biomechanics of the lateral ligaments of the ankle: an evaluation of the effects of axial load and single plane motions on ligament strain patterns, *Foot Ankle*, 12, 92, 1991.

13. de Lange, A., Huiskes, R., and Kauer, J.M., Effects of data smoothing on the reconstruction of helical axis parameters in human joint kinematics, *J. Biomechanical Eng.*, 112, 197, 1990.

14. de Lange, A., Huiskes, R., and Kauer, J.M., Measurement errors in roentgen-stereophotogrammetric joint-motion analysis, *J. Biomechanics*, 23, 259, 1990.

15. de Lange, A., Huiskes, R., and Kauer, J.M., Wrist-joint ligament length changes in flexion and deviation of the hand: an experimental study, *J. Orthop. Res.*, 8, 722, 1990.

16. Erickson, A.R., Yasuda, K., Beynnon, B., Johnson, R., and Pope, M., An *in vitro* dynamic evaluation of prophylactic knee braces during lateral impact loading, *Am. J. Sports Med.*, 21, 26, 1993.

17. Fung, Y.C., Biomechanics: its scope, history, and some problems of continuum mechanics in physiology, *Appl. Mech. Rev.*, 21, 1, 1967.

18. Fung, Y.C., Elasticity of soft tissues in simple elongation, *Am. J. Physiol.*, 213, 1532, 1967.

18a. Fung, Y.C., *Biomechanics Mechanical Properties of Living Tissues*, Springer-Verlag, NY, 1981.

19. Haut, R.C. and Little, R.W., A constitutive equation for collagen fibers, *J. Biomechanics*, 5, 423, 1972.

20. Huiskes, R., Kremers, J., Lange, A., de Woltring, H.J., Selvik, G., and van Rens, T.J.G., An analytical stereophotogrammetric method to determine the three-dimensional geometry of articular surfaces, *J. Biomechanics*, 18, 559, 1985.

21. Kristal, P., Tencer, A.F., Trumble, T.E., North, E., and Parvin, D., A method for measuring tension in small ligaments: an application to the ligaments of the wrist carpus, *J. Biomechanical Eng.*, 115, 218, 1993.

22. Lewis, J.L., Lew, W.D., and Schmidt, J., A note on the application and evaluation of the buckle transducer for the knee ligament force measurement, *J. Biomechanical Eng.*, 104, 125, 1982.

23. Meglan, D., Berme, N., and Zuelzer, W., On the construction, circuitry and properties of liquid metal strain gages, *J. Biomechanics*, 21, 681, 1988.

23a. Mow, V.C. and Hayes, W.C., Eds., *Basic Orthopaedic Biomechanics*, Raven Press, NY, 1991.

23b. Microstrain, An information brochure, 294 North Winooski Ave., Burlington VT.

24. Noyes, F.R., Butler, D.L., Grood, E.S., Zernicke, R.F., and Hefzy, M.S., Biomechanical analysis of human ligament grafts used in knee ligament repairs and reconstructions, *J. Bone Jt. Surg.*, 66A, 344, 1984.

25. Noyes, R.F., DeLucas, J.L., and Torvik, P.J., Biomechanics of anterior cruciate ligament failure: an analysis of strain rate sensitivity and mechanisms of failure in primates, *J. Bone Jt. Surg.*, 56, 236, 1974.

26. Renstrom, P., Wertz, M., Incavo, S., Pope, M., Ostgaard, H.C., Arms, S., and Haugh, L., Strain in the lateral ligaments of the ankle, *Foot Ankle*, 9, 59, 1988.

27. Selvik, G., Roentgen stereophotogrammetry: a method for the study of the kinematics of the skeletal system, *Acta Orthop. Scand.* (suppl.), 232, 1, 1989.

28. Selvik, G., Roentgen stereophotogrammetric analysis, *Acta Radiologica*, 31, 113, 1990.

29. Smith, A., The diagnosis and treatment of injuries to the cruciate ligaments, *Br. J. Surg.*, 6, 179, 1918.

30. Stone, J.E., Madsen, M.H., Milton, J.L., Swinson, W.F., and Turner, J.L., Developments in the design and use of liquid-metal strain gages (biomechanics applications), *Exp. Mechanics*, 23, 129, 1983.

31. Viidik, A., Properties of tendons and ligaments, in *Handbook of Bioengineering*, Skalak, R. and Chien, S., Eds., McGraw-Hill, New York, 1987.

32. Viidik, A., Structure and function of normal and healing tendons and ligaments, in *Biomechanics of Diarthrodial Joints*, Mow, V.C., Ratcliffe, A., and Woo, S. L.-Y., Eds., Springer-Verlag, New York, 1990, vol. 1.

33. Warren, L.A., Marshall, J.L., and Girgis, F., The prime static stabilizer of the medial side of the knee, *J. Bone Jt. Surg.*, 56A, 665, 1974.

34. Weaver, L., Tencer, A.F., and Trumble, T.E., Tensions in the palmar ligaments of the wrist I. The normal wrist, *J. Hand Surg.*, 19A, 464, 1994.

35. Woltring, H.J., Huiskes, R., and Veldpaus, F.E., Finite centroid and helical axis estimation from noisy landmark measurements in the study of human joint kinematics, *J. Biomechanics*, 18, 379, 1985.

36. Woo, S.L., Danto, M.I., Ohland, K.J., Lee, T.Q., and Newton, P.O., The use of a laser micrometer system to determine the cross-sectional shape and area of ligaments: a comparative study of two existing methods, *J. Biomechanical Eng.*, 112, 426, 1990.

Index